ANALYTICAL CHEMISTRY AND MICROCHEMISTRY

PERVAPORATION

PROCESS, MATERIALS AND APPLICATIONS

Analytical Chemistry and Microchemistry

Additional books and e-books in this series can be found on Nova's website under the Series tab.

ANALYTICAL CHEMISTRY AND MICROCHEMISTRY

PERVAPORATION

PROCESS, MATERIALS AND APPLICATIONS

JEAN GARCIA
EDITOR

nova science publishers
New York

Library of Congress Cataloging-in-Publication Data

ISBN: 978-1-53614-459-8

Published by Nova Science Publishers, Inc. † New York

CONTENTS

PREFACE

Pervaporation is a separation process in which the selective permeation of components of a liquid mixture is achieved by way of a chemical potential gradient through a non-porous membrane. In Pervaporation: Process, Materials and Applications, the fundamentals and applications of pervaporation are described as a promising technique for the recovery of flavor compounds from dilute aqueous solutions, separation of azeotropic mixtures and for the dehydration of organic solvents.

This collection also describes history of pervaporation in an effort to outline the differences between this and other membrane separation technologies including dialysis, ultrafiltration, microfiltration, nanofiltration and reverse osmosis.

The closing chapter focuses on the authors' on-going development of high performance bio-based cellulosic membranes for ethyl tert-butyl ether purification by pervaporation. Cellulose acetate is extremely selective for ethanol removal from ethyl tert-butyl ether, however its flux is very low. Different strategies for improving its flux while maintaining a high selectivity are described and the main relationships between membrane structure, morphology and properties are illustrated.

Chapter 1 - Pervaporation is a separation process in which the selective permeation of components of a liquid mixture is achieved due to a chemical potential gradient through a non-porous membrane. In this chapter the fundamentals and applications of pervaporation are described as a promising technique for the recovery of flavor compounds from dilute aqueous solutions, separation of azeotropic mixtures and also for the dehydration of organic solvents. Aspects related to the mechanism of mass transport through the pervaporation membranes are briefly discussed. Details of pervaporation applications focusing on the efficiency of the separation process, types of membranes, manufacturing procedures and the recent development of membranes blended with natural biopolymers are also discussed. As a perspective of the pervaporation separation process, the chapter includes the integrated processes of pervaporation with other separation or reaction systems, such as membrane reactors, and their applications.

Chapter 2 - In this chapter, at first the definition, history and characteristics of pervaporation (PV) are described to understand the differences with other membrane separation technologies such as dialysis, ultrafiltration, microfiltration, nanofiltration and reverse osmosis. Furthermore, PV principle and PV fundamental analysis such as permeation equation are described. It is very important that we understand them in a judgment when the author's apply the membrane which the author's developed to the real application field. In practical fields PV membrane materials: hydrophilic polymer materials, hydrophilic inorganic materials, hydrophobic polymer membrane materials, water permselective and organic permselective composite mem-branes, water-permselective, organic-permselective and organic-organic separation hybrid membranes, inorganic membranes, pervaporation technol-ogies: dehydration such as water-alcohol dehydration and water-organic dehydration, organic separation from water: alcohol-selective concentration and organic selective removal, organic mixture separation such as separation of alcohol isomer, xylene mixtures, ether-alcohol mixture, hydrocarbon isomer, organic-organic azeotrope and desalination by PV, and technologies of wide field in PV are discussed in detail.

Chapter 3 - Ethyl *tert*-butyl ether (ETBE) is a major biofuel mainly produced from the reaction of isobutene with a large excess of bioethanol. This ether greatly improves fuel combustion and limits the emission of toxic hydrocarbons. Moreover, thanks to its particular molecular structure, this ether does not accumulate into the environment contrary to methyl *tert*-butyl ether (MTBE). Nevertheless, during the industrial ETBE synthesis, this ether forms an azeotropic mixture containing 20 wt% of ethanol, which must be removed during the final purification stage. The ternary distillation process generally used for this separation is highly energy intensive. Several works have shown that the pervaporation membrane process (PV) could be a very good alternative for ETBE purification with very important energy savings. This chapter focuses on our on-going development of high performance bio-based cellulosic membranes for ETBE purification by PV. Cellulose acetate (CA) is extremely selective for ethanol removal from ETBE but its flux is very low. Different strategies for greatly improving its flux while maintaining a very high selectivity are described and the main relationships between membrane structure, morphology and properties are illustrated. In a first part, the grafting of CA by controlled radical polymerization enables to vary the length and the number of polymethacrylate grafts, inducing different copolymer morphologies and PV properties. In particular, it is shown that, for a same graft content in the bio-based membranes, the nanostructuration of the grafts plays a determining role on the separation features. The second part explores the development of purely bio-based membranes made of a CA backbone with bio-based polylactide grafts based on a controlled grafting by "click" chemistry. This strategy enables to plasticize the cellulosic membranes very efficiently and greatly improves the membrane properties. In the third part, the grafting of ionic liquids onto CA is reported as another very efficient way of inducing CA plasticization. The influence of the ionic liquid content and molecular structure is discussed to reveal the key parameters for the membrane properties. Several bio-based membranes with grafted ionic liquids have very improved pervaporation flux (up to almost 20 times that of virgin CA) and permeate ethanol only, corresponding to outstanding infinite separation factors for this challenging liquid/liquid separation.

In: Pervaporation
Editor: Jean Garcia
ISBN: 978-1-53614-459-8

Chapter 1

PERVAPORATION: FUNDAMENTALS AND APPLICATIONS

Adriane Bianchi Pedroni Medeiros[1,*], Joel Robert Karp[3], Vitor Renan da Silva[2], Suzan Cristina Rossi[1] and Moisés A. Marcelino Neto[3]

[1]Department of Bioprocess Engineering and Biotechnology, Federal University of Paraná, Curitiba, Brazil
[2]Department of Chemical Engineering, Federal University of Paraná, Curitiba, Paraná, Brazil
[3]Department of Mechanical Engineering, Federal University of Technology, Paraná, Curitiba, Brazil

ABSTRACT

Pervaporation is a separation process in which the selective permeation of components of a liquid mixture is achieved due to a chemical potential gradient through a non-porous membrane. In this chapter the fundamentals and applications of pervaporation are described as a promising technique for the recovery of flavor compounds from dilute aqueous solutions, separation of azeotropic mixtures and also for the

[*] Corresponding Author Email: adrianebpm@ufpr.br.

dehydration of organic solvents. Aspects related to the mechanism of mass transport through the pervaporation membranes are briefly discussed. Details of pervaporation applications focusing on the efficiency of the separation process, types of membranes, manufacturing procedures and the recent development of membranes blended with natural biopolymers are also discussed. As a perspective of the pervaporation separation process, the chapter includes the integrated processes of pervaporation with other separation or reaction systems, such as membrane reactors, and their applications.

Keywords: separation process, membrane, mass diffusion, solution-diffusion

1. Introduction

Pervaporation is a membrane separation process in which the selective permeation of components of a liquid mixture is achieved due to a chemical potential gradient through a non-porous membrane, promoting the diffusion, normally designated as permeation, of specified compounds followed by the partial evaporation on the permeate stream. The selectivity of the process is due to the fact that different species permeate through the membrane at different rates, according to their structure, chemical affinity among other parameters. Therefore, molecules which are easily permeated through the membrane can be obtained on the permeate side at high concentration, even when from a very dilute liquid feed solution.

The concept of pervaporation was emerged within the scientific community at the beginning of the 20th century with sporadic studies of permeation correlated phenomena. The definition of the pervaporation process was firstly introduced by Kober (1917) in the study "Pervaporation, perstillation and percrystalization", where a selective permeation of water from a feed mixture of water/toluene through a natural membrane of "collodion" was evaluated. After this initial identification of the afore-mentioned phenomena, the first major research effort specifically focused on pervaporation was registered in the late 1950's by Binning and associates, which developed studies of selective separation by membrane pervapo-ration. This innovative work presented at the 1958 American Chemical Society Meeting exposed the applicability of pervaporation in the processing of a ternary azeotropic mixture of isopropanol/ethanol/water overhead of a classic distillation column. Heissler et al. (1956) also pioneered on this field, promoting an effective dehydration of ethanol employing a regenerated cellulose film.

The aforementioned researchers greatly contributed to this field, introducing a new procedure applicable to a wide range of industrial aims. However, the conducted research exposed one major restriction regarding this topic, which was the low magnitude of the permeation flow rate. Therefore, further application of this process in an industrial scale would require very large permeation areas, restricting the use of pervaporation from an economical point of view. This represented a major encouragement for future researchers at the time, since an improvement in the permeation flux would allow the employment of this highly efficient and selective process in an industrial scale.

To this aim, Loeb and Sourirajan in the year of 1962 developed the phase-inversion technique, which allowed the synthesis of asymmetric membranes, which consists in a very thin dense membrane layer, placed over a porous support layer. The lower thickness of the non-porous membrane achieved through this technique yielded a larger permeation flux, maintaining the high efficiency and selectivity obtained earlier.

The process of pervaporation represents even nowadays an interesting alternative for a wide range of applications in industry, where the high selectivity and efficiency can hardly be obtained by traditional processes. In this scenario, pervaporation remains still a relevant subject within the scientific community, with many qualified studies being conducted up to the present date. The present chapter aims to expose the main aspects of pervaporation, focusing on the fundamentals, processes as well as new applications regarding this technique.

2. Fundamentals of Pervaporation

2.1. The Solution-Diffusion Model

Pervaporation is a membrane process which is used to separate a liquid mixture by a selective diffusion through a non-porous membrane followed by a partial vaporization on the permeate side. Although other experimental setups may be employed in pervaporation experiments, the most commonly applied system is the one shown in Figure 2.1., which consists in a continuous vacuum-drive pervaporation system. The liquid mixture is allowed to flow along one side of the membrane and a fraction of it suffers diffusion and is vaporized on the opposite side of the membrane. This permeate stream is collected in the liquid state after condensation on a condenser system, usually employing liquid nitrogen. The driving force in the system shown in Figure 2.1. is the chemical potential gradient, promoted by an intense pressure difference on the two sides of the membrane.

Due to the presence of the membrane, the pervaporation process summarizes the permeation and partial evaporation steps, being therefore much more complex than traditional separation procedures based entirely on liquid-vapor equilibrium, such as distillation. This increased complexity leads to a more elaborated phenomenological modelling, so one can easily expect that the permeate composition must be differently estimated in comparison to the vapor phase obtained after the establishment of a free liquid-vapor equilibrium, in which the interactions with the membrane are not present. The classical approach used to model the pervaporation phenomena is denominated solution-diffusion, schematically shown in Figure 2.2. This approach has been widely applied in pervaporation assays, stating that the overall pervaporation process can be described in three successive steps (Huang and Rhim, 1991):

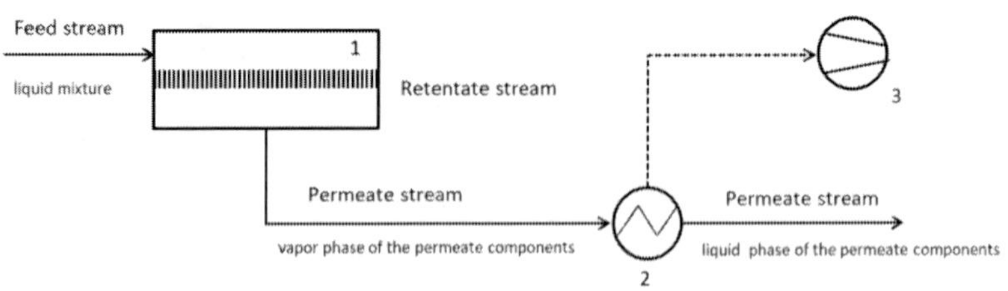

Figure 2.1. Continuous-flow of pervaporation unit. 1 – Membrane unit; 2 – condenser system; 3 – vacuum pump.

a) Upstream partitioning of the feed components between the flowing liquid feed stream and the swollen upstream surface layer of the membrane (sorption of the selective components by surface membrane);

b) Internal diffusion of the components from the surface of the membrane through the internal free volume fraction of the nonporous layer;

c) Desorption of the permeated components at the downstream surface of the membrane.

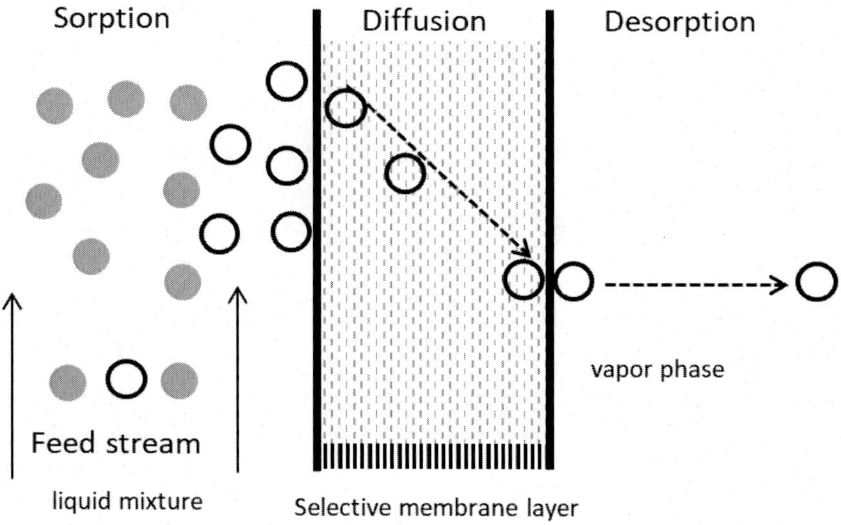

Figure 2.2. Schematic diagram of the solution-diffusion model applied for pervaporation. ○ Component with affinity to the membrane surface. ● Non-interacting component.

The sorption and diffusion in the nonporous membrane are normally considered as the governing steps in the permeation process, since the desorption stage consists mainly in the evaporation of the component adsorbed onto the downstream surface of the membrane after the diffusion stage. This step occurs rapidly in comparison with the other previous stages, in a way that all the components permeated will easily be evaporated in the interface of the membrane. Therefore, it is considered that the selectivity of the pervaporation process is dependent only of the sorption and diffusion steps.

Both of these stages influence the overall permeation process and the understanding of these individual steps relies on several parameters. For instance, regarding the selective sorption step, thermodynamic aspects of the liquid-solid equilibrium greatly influences the solubility of the components in the membrane and, therefore, the sorption effect. The diffusion step, on the other hand, is more affected by other parameters more directly related to mass transfer mechanisms, like the chemical structure of the components as well as membrane properties such as morphology and composition. Structural aspects of the membrane are also of great relevance, such as the segmental mobility and free volume fraction. The evaluation of the individual influence of each of these two steps is a difficult task. However, based on the aforementioned parameters, it can be said that generally a selective sorption of the components present in the liquid feed solution is carried out according to the solid-liquid equilibrium present, in a way that the sorption selectivity is related to the thermodynamic interactions of the individual compounds with the membranes interface.

The diffusion step is also responsible for a selective permeation of the compounds, with this selectivity being mainly due to different diffusion rates across the membrane. Therefore, more permeable components are more easily diffused through the membrane, in comparison with a lesser permeable component. Although this stage is also related to thermodynamic interactions with the membrane, it is understandable that the components structure and chemical affinity are more crucial in the diffusion stage within the free volume fraction of the non-porous layer.

2.2. The Permeation Process

This chapter presents a modelling primarily based on the solution-diffusion model. Several assumptions have to be carried out regarding the phenomenological aspects of the pervaporation phenomena. Those hypothesis approach the aforementioned thermodynamic, chemical as well as structural aspects, in order to adequately describe the phenomena and present a thorough modelling. These following assumptions were carried out:

1. The vapor phase can be approached as an ideal gas;
2. The liquid mixture in the feed stream presents incompressible behaviour;
3. The pressure and temperature across the membrane are constant and equal to the feed stream;
4. The activity coefficient inside the membrane is constant;
5. The diffusion coefficient is constant across the thickness of the membrane;
6. An equilibrium state between the upside surface of the membrane and the liquid phase (feed stream) is assumed;
7. An equilibrium state between the downside surface of the membrane and the vapor phase (permeate stream) is assumed.

The driving force that induces the mass transfer during the pervaporation process is generally recognized as a chemical potential gradient across the membrane. This chemical potential (μi) can be expressed in terms of the composition, pressure and temperature according to Eq. 1(Sandler, 1999):

$$\mu_i = \mu_i^O + R \cdot T \cdot ln[a_i] + \int_{P_0}^{P} V_i^M dP - \int_{T_0}^{T} S_i \cdot dT \tag{1}$$

where: μ_i^O is the standard chemical potential, a_i is the activity of the i component, R is the universal gas constant, T is the absolute temperature, P is the pressure, V_i^M is the molar volume of the i component and S_i is the entropy of the i component.

Since the pervaporation process depends greatly in the diffusion stage across the membrane, it is clear that the permeation has to be accompanied by a chemical potential gradient between the upstream and downstream interfaces of the membrane. At this point, researchers usually distinguish three different systems of pervaporation units, each one differing from the other regarding the origin of the chemical potential gradient. As Eq. 1 clearly exposes, a chemical potential gradient in the system can be generated by altering the composition, temperature or pressure of the system. Therefore, each pervaporation system is designed in order to properly manipulate one of the aforementioned variables, resulting in an adequate chemical potential gradient. Among the pervaporation systems frequently employed, the vacuum-driven pervaporation (Figure 2.3 – A) is the most commonly applied pervaporation mode, in which the desired chemical potential gradient is easily reached by employment of an intense vacuum on the permeate side. Figure 2.4 shows a general illustration of the chemical potential gradient in a vacuum-driven pervaporation setup. By making an analogy with the modified Raoult's law (where the vapor phase is an ideal gas but the liquid phase is related to its activity coefficient) the mass transfer throughout the membrane is carried out by maintaining the partial pressure of the gas on the permeate side of the membrane lower than the vapor pressure of the same component in the feed liquid, multiplied by its activity coefficient and molar fraction.

The chemical potential gradient also can be obtained by a reduction of the partial pressure of the permeate stream by using a purge gas (Figure 2.3 – B) or a temperature difference between the feed and the permeate streams (Figure 2.3 – C) by use of a heat exchanger before the pervaporation unit.

The diffusion across the membrane in the pervaporation process is usually described by the Fick's law of diffusion, which can be more generally written according to Eq. 2, where $J_i(x)$ is the molar flux of the i component in a known location in the membrane; L_i is the mass transfer coefficient of the i component.

It is necessary to describe the permeation flux by measurable and more feasible variables. As previously exposed, the chemical potential gradient depends on the pressure, temperature and composition of the system. Regardless of the pervaporation system and chemical potential gradient employed, the mass transfer process across the pervaporation membrane usually occurs at a small membrane thickness, so that the temperature variation is negligible. Therefore, pervaporation can be dealt with as an isothermal process across the membrane, where the temperature is equal to the temperature of the feed stream. Also, hypothesis 3 states that the pressure across the membrane is constant and equal to the pressure of the liquid feed stream. This assumption follows from the fact that the desorption stage, that is, the evaporation of the components after the selective sorption and diffusion steps, occurs only at the interface with the permeate side. This means that the two major stages regarding the membranes selectivity occur at liquid phase. Therefore, from the incompressible fluid assumption, there should be no reason to considerer some pressure loss across the membrane, with the pressure gradient abruptly appearing at the membrane/permeate interface, as Figure 2.4. exposed.

Therefore, from Eq. 1, the chemical potential gradient can be written in the differential form by evaluating only the difference in composition, according to Eq. 3:

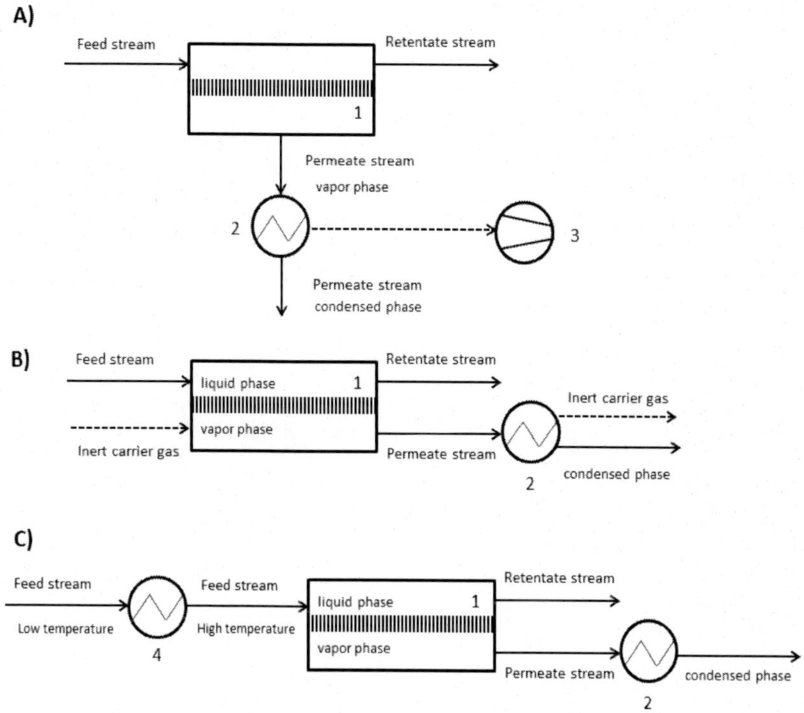

Figure 2.3. Usual pervaporation modes of operation. A) Vacuum-driven pervaporation. B) Carrier gas pervaporation. C) Temperature-driven pervaporation. 1 – Pervaporation membrane unit. 2 – Condenser system. 3 – Vacuum pump system. 4 – Heat exchanger.

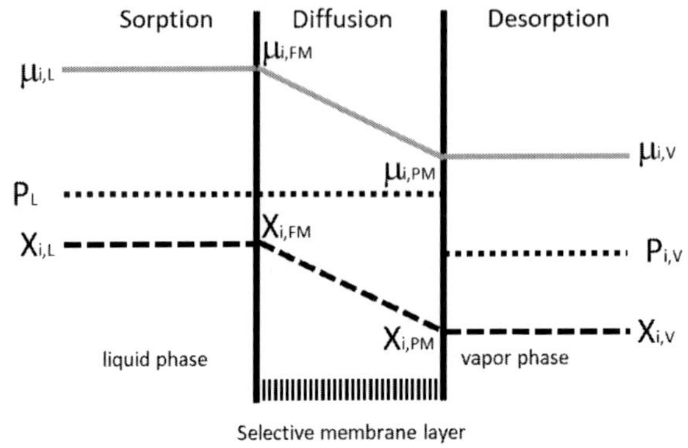

Figure 2.4. Chemical potential, pressure and molar fraction profiles through a pervaporation membrane following the solution-diffusion model (Adapted from Wijmans and Baker, 2006).

$$J_i(x) = -L_i \cdot \frac{d\mu_i}{dx} \tag{2}$$

$$d\mu_i \equiv R \cdot T \cdot \frac{d\, a_i}{a_i} \tag{3}$$

By equations (2) and (3), the permeate flux can be described by Eq. 4:

$$J_i(x) = \frac{-L_i \cdot R \cdot T}{a_i} \cdot \frac{da_i}{dx} \tag{4}$$

Since the i component remains liquid throughout the thickness of the membrane, the activity is described by Eq. 5 and the molar permeation flux of the i component can be rewritten in terms of the activity coefficient, γ_i, and molar fraction, X_i, of the component i, according to Eq. 6.

$$a_i \equiv X_i \cdot \gamma_i \tag{5}$$

$$J_i(x) = \frac{-L_i \cdot R \cdot T}{X_i \cdot \gamma_i} \cdot \frac{d(X_i \cdot \gamma_i)}{dx} \tag{6}$$

As related by Baker and Wijmans (1995), the gradient of activity along a dense membrane in a pervaporation process is smooth, approximately constant. That is, the activity gradient across the membrane is only due to the molar fraction variation, with the activity coefficient considered constant. This fact is supported by most of the research conducted in the field, which express the molar permeation flux as a function of the molar fraction alone, according to Eq. 7:

$$J_i(x) = \frac{-L_i \cdot R \cdot T}{X_i} \cdot \frac{dX_i}{dx} \tag{7}$$

This equation has the same form as the Fick's law of diffusion in a stationary medium. Therefore, the diffusion coefficient of the i component, D_i, can be defined according to Eq. 8 and the permeation flux through the membrane can be rewritten, as exposed by Eq. 9.

$$D_i = \frac{L_i \cdot R \cdot T}{X_i} \tag{8}$$

$$J_i(x) = -D_i \cdot \frac{dX_i}{dx} \tag{9}$$

The approach presented in this chapter, based on the solution-diffusion model, still considers that the diffusion coefficient remains constant during the pervaporation under a stationary regime. Therefore, the average permeation flux can be obtained by simple integration of Eq. 9 with the boundary conditions in the membrane interfaces illustrated in Figure 2.4, resulting in Eq. 10.

$$J_i = D_i \cdot \frac{X_{i,FM} - X_{i,PM}}{\lambda_M} \tag{10}$$

where: λ_M is the membrane thickness, $X_{i,FM}$ is the molar fraction of the i component in the upside surface of the membrane in contact with feed stream; and $X_{i,PM}$ is the molar fraction of the i component in the downside surface of the membrane in contact with permeate stream.

Despite the fact that molar concentrations on the interfaces of the membrane are not easily measured, Eq. 10 is of great importance since it is more easily converted into measurable variables. This can be achieved by a manipulation of variables based on the boundary conditions of the problem. The assumption of phase equilibrium at the interface of the membrane with the feed solution yields an equality of the chemical potential at the interface of the membrane, $\mu_{i,FM}$, with the chemical potential in the liquid feed solution, $\mu_{i,L}$. Similarly, for the interface of the membrane with the permeate side, the chemical potential at the membrane, $\mu_{i,PM}$, equals to the chemical potential of the vapor permeate stream, $\mu_{i,P}$. It follows from Eq. 1, that the chemical potential in an incompressible liquid can be described by Eq. 11, considering an isothermal process and the saturation condition as a reference state. In Eq. 11, P_F denotes the pressure on the liquid feed stream. Similarly, for the vapor phase, where an ideal gas behavior is assumed (in the case of vacuum-driven pervaporation, it should be remembered that an intense vacuum is applied on the permeate side), Eq. 12 quantifies the chemical potential of this phase, under the same assumptions.

$$\mu_{i,L} = \mu_i^{sat} + R \cdot T \cdot ln\left[\gamma_{i,L} \cdot X_{i,L}\right] + V_i^M \cdot \left(P_F - P_i^{sat}\right) \tag{11}$$

$$\mu_{i,V} = \mu_i^{sat} + R \cdot T \cdot ln\left[\gamma_{i,P} \cdot X_{i,P}\right] + R \cdot T \cdot ln\left[\frac{P_P}{P_i^{sat}}\right] \tag{12}$$

Still under the assumption of phase equilibrium at the interfaces of the membrane and the consequent chemical potential equality with the bulk solution, it must be pointed out that, before the desorption stage, only a liquid phase is present. Therefore, the equilibrium equations at the interface of the membrane with the permeate side is described by Eqs. 13 and 14.

$$\mu_{i,PM} \equiv \mu_{i,P} \tag{13}$$

$$\mu_i^{sat} + R \cdot T \cdot ln\left[\gamma_{i,P} \cdot X_{i,P}\right] + R \cdot T \cdot ln\left[\frac{P_P}{P_i^{VAP}}\right] = \mu_i^{sat} + R \cdot T \cdot ln\left[\gamma_{i,PM} \cdot X_{i,PM}\right] + V_i^M \cdot \left(P_F - P_i^{sat}\right) \tag{14}$$

Rearranging Eq. 14:.

$$\gamma_{i,P} \cdot X_{i,P} \cdot \left[\frac{P_P}{P_i^{sat}}\right] = \gamma_{i,PM} \cdot X_{i,PM} \cdot exp\left[\frac{V_i^M}{R \cdot T} \cdot \left(P_F - P_i^{sat}\right)\right] \tag{15}$$

The last term in Eq. 15 is known also as the Poynting correction factor, which is usually negligible at low to moderate pressures. The pervaporation process, however, usually operates at low pressures on both sides, with the pressure on the feed stream being close to the atmospheric. Therefore, the last term in Eq. 15 approaches the unity (Koretsky, 2007). For an ideal gas mixture, the product between molar fraction of the i component in permeate stream, $X_{i,P}$, and vapor phase pressure, P_P, is defined as the partial pressure of the i component, $P_{i,P}$, in the permeate stream (Sandler, 1999). Therefore, the molar fraction of the i component at the interface of the membrane on the permeate side can be expressed by Eq. 16.

$$X_{i,PM} = \frac{\gamma_{i,P} \cdot P_{i,P}}{P_i^{sat} \cdot \gamma_{i,PM}} \tag{16}$$

Baker and Wijmans (1995) define the sorption coefficient of the species i in the gas phase, K_i^G, according to Eq. 17, which correlates the composition at the interface of the downside surface membrane with the adjacent vapor permeate stream.

$$K_i^G = \frac{\gamma_{i,P}}{P_i^{sat} \cdot \gamma_{i,PM}} \tag{17}$$

The molar fraction of the i component in the downside surface of the membrane can, therefore, be determined by Eq. 18, that is, in terms of the partial pressure of the component of interest and the sorption coefficient in the gas phase.

$$X_{i,PM} = K_i^G \cdot P_{i,P} \tag{18}$$

The same approach should be used to evaluate the molar fraction of the i component in the upside surface of the membrane, $X_{i,FM}$, which is in contact with the feed stream. The chemical potential in the upside surface of the membrane, $\mu_{i,F}$ equals the chemical potential in the feed stream, $\mu_{i,FM}$, as exposed by Eq. 19. Substituting Eq. 11 (applied both to the feed and solution and the membrane interface) into Eq. 19, gives the phase equilibrium described by Eq. 20.

$$\mu_{i,F} \equiv \mu_{i,FM} \tag{19}$$

$$\mu_i^{sat} + R \cdot T \cdot ln\left[\gamma_{i,F} \cdot X_{i,F}\right] + V_i^M \cdot \left(P_F - P_i^{sat}\right) = \mu_i^{sat} + R \cdot T \cdot ln\left[\gamma_{i,FM} \cdot X_{i,FM}\right] + V_i^M \cdot \left(P_F - P_i^{sat}\right) \tag{20}$$

Following the same procedure for the vapor permeate stream, a sorption coefficient related to the liquid phase, K_i^L, is defined by Eq. 22. Therefore, the molar fraction at the interface of the membrane with the feed solution can be related to its composition in the bulk solution and a sorption coefficient in the liquid phase, as described by Eq. 23. It should be noted that this relation of the molar composition at the membranes interface is similar with the downside surface of the membrane, where the composition is adequately described by the partial pressure.

$$X_{i,FM} = \frac{\gamma_{i,F} \cdot X_{i,F}}{\gamma_{i,FM}} \tag{21}$$

$$K_i^L = \frac{\gamma_{i,P}}{\gamma_{i,FM}} \tag{22}$$

$$X_{i,FM} = K_i^L \cdot X_{i,F} \tag{23}$$

At this point, the average permeation flux of the i component (J_i) can be rewritten by substituting Eqs. 18 and 23 in Eq. 10, leading to Eq. 24. This equation clarifies that an important step towards the phenomenological modeling of the permeation flux has been achieved, since the permeation flux is now presented as a function of the molar fractions at the bulk of the solutions, with the knowledge of the composition at the interface of the membrane being no longer necessary.

$$J_i = D_i \cdot \frac{K_i^L \cdot X_{i,F} - K_i^G \cdot P_{i,P}}{\lambda_M} \tag{24}$$

However, further improvement of Eq. 24 is still required, since it is desirable to evaluate the contribution of the sorption stage at only one phase, liquid or gaseous. As will be further demonstrated, the employment of only one sorption coefficient will allow the definition of one important parameter in the pervaporation phenomena, the permeability. At this point, therefore, it is convenient to represent the influence of the sorption stage in the gas phase only. This can be done by the definition of a hypothetic gas phase (denoted by the superscript G), which is in equilibrium with the feed solution, so that the chemical potential of these two phases are equal. Therefore, the chemical equilibrium at the interface of the membrane with the feed solution, given by Eq. 19, can be rewritten in terms of this hypothetic gas phase, as Eq. 25 indicated. The chemical potential of this arbitrary gas phase and the governing equation of the aforementioned phase equilibrium are given, respectively, by Eqs. 26 and 27.

$$\mu_{i,F} \equiv \mu_{i,F}^{G} = \mu_{i,FM} \tag{25}$$

$$\mu_{i,F}^{G} = \mu_i^{sat} + R \cdot T \cdot ln\left[\gamma_{i,F}^{G} \cdot X_{i,F}^{G}\right] + R \cdot T \cdot ln\left[\frac{P_F}{P_i^{sat}}\right] \tag{26}$$

$$\mu_i^{sat} + R \cdot T \cdot ln\left[\gamma_{i,F} \cdot X_{i,F}\right] + V_i^M \cdot \left(P_F - P_i^{sat}\right) = \mu_i^{sat} + R \cdot T \cdot ln\left[\gamma_{i,F}^{G} \cdot X_{i,F}^{G}\right] + R \cdot T \cdot ln\left[\frac{P_F}{P_i^{sat}}\right] \tag{27}$$

The composition of the feed liquid mixture can be expressed in terms of the hypothetic gas phase, according to Eq. 28. Therefore, from this equilibrium analysis, Eq. 23 can be combined with Eqs. 22, 28 and 17, yielding the molar fraction of the i component at the membranes interface with the feed solution as a function of the sorption coefficient in the gas phase, as given by Eq. 29.

$$X_{i,F}^{L} = \frac{\gamma_{i,F}^{G} \cdot P_{i,F}^{G}}{\gamma_{i,F}^{L} \cdot P_i^{sat}} \tag{28}$$

$$X_{i,FM} = K_i^G \cdot P_{i,F} \tag{29}$$

With the above mathematical procedure, Eq. 24 can be improved so that the molar permeation flux can be estimated from a sorption coefficient of the gas phase only, yielding Eq. 30. Since the approach presented in this chapter is primarily based on the solution-diffusion model, it is quite comprehensible that the selectivity on the pervaporation process depends of the individual diffusion and sorption stages, represented in Eq. 30, respectively, by the diffusivity and sorption coefficient at the gas phase. At this point, an important definition is made, by finding a single parameter that involves both of this stages. This parameter, called permeability, Q_i^G, is of great importance in all membrane separation processes. It is firstly defined for the gas phase as the product of the diffusivity and sorption coefficient, according to Eq. 31. This important parameter represents a single coefficient of proportionality, related to both diffusion and sorption individual steps, being, therefore, related to the global selectivity of the pervaporation process. The molar permeation flux can, therefore, be expressed according to Eq. 32, in terms of the permeability in the gas phase across the thickness of the membrane and the difference in partial pressure of the i component.

$$J_i = D_i \cdot K_i^G \cdot \frac{p_{i,F} - p_{i,P}}{\lambda_M} \tag{30}$$

$$Q_i^G = D_i \cdot K_i^G \tag{31}$$

$$J_i = Q_i^G \cdot \frac{p_{i,F} - p_{i,P}}{\lambda_M} \tag{32}$$

Equation 32 is analogous to other constitutive equations related to transport phenomena processes, such as the Fourier law for heat transfer and the Fick's law of diffusion. The molar permeation flux can, therefore, be expressed in term of a driving force (difference in partial pressure) divided by a resistance term, related to the thickness of the membrane and the permeability of the i component. The main difference from the original Fick's law of diffusion is that this resistance term is due not only to the diffusion across the membrane, but to the sorption stage also. Eq. 32 represents an important advancement since the molar permeation flux is estimated according to measurable variables, that is, the pressure on the permeate and feed streams. However, Eq. 32 is more frequently employed for gas permeation problems, in which both phases are gaseous. Although the applicability of this equation in pervaporation assays is acceptable and has been carried out over the years, its use requires more complex calculation, since in pervaporation applications the feed solution is liquid. The estimation of the partial pressure on the stream side remains straight forward, however, the determination of the partial pressure on the feed stream is more complex. This is normally carried out via computer simulations and the selection of a proper equation of state. Also, for the sake of completion, it is desirable to rearrange Eq. 32, in order to relate the molar permeation flux to a permeability coefficient in the liquid phase. Substituting Eq. 28 into Eq. 32:

$$J_i = \frac{Q_i^G}{\lambda_M} \cdot \left(\frac{\gamma_{i,F}^L \cdot X_{i,F}^L \cdot P_i^{sat}}{\gamma_{i,F}^G} - P_{i,P} \right) \tag{33}$$

Baker and Wijmans (1995) reported also that the Henry coefficient in terms of volatility (in pressure units) can be estimated according to Eq. 34. This coefficient is normally employed to compare the composition of gas and liquid phases under infinite dilution equilibrium. Therefore, the molar permeation flux can be written in terms of the feed liquid composition and Henry coefficient, as indicated by Eq. 35.

$$H_i = \frac{\gamma_{i,F}^L \cdot P_i^{sat}}{\gamma_{i,F}^G} \tag{34}$$

$$J_i = \frac{Q_i^G}{\lambda_M} \cdot \left(H_i \cdot X_{i,F}^L - P_{i,P} \right) \tag{35}$$

By mathematical manipulation, the sorption coefficient in the liquid phase given by Eq. 22 may also be expressed in terms of the sorption coefficient in the gas phase and the Henry constant, by substituting Eqs. 17 and 34. Therefore, from the definition of permeability given Eq. 31, the

Henry constant can be employed in the relation between the permeability coefficient in the gas and liquid phases, as exposed in Eq. 37.

$$K_i^L = K_i^G H_i \tag{36}$$

$$Q_i^L = Q_i^G H_i \tag{37}$$

Finally, substituting Eq. 37 in Eq. 35, the molar permeation flux of the i component can be estimated according to Eq. 38. The approach of the solution-diffusion model presented in this chapter yields a governing equation for the molar permeation flux that is more adequate to pervaporation applications, since the driving force term is related to the molar composition of the liquid phase and the partial pressure in the permeate stream. Pervaporation assays can, therefore, be more adequately described by this equation, since the feed solution and the permeate stream are, respectively, in the liquid and gas phase. In addition, the proportionality coefficient in Eq. 38 is no longer the permeability in the gas phase, but considerers that the both individual steps, sorption and diffusion across the membrane, occur in the liquid phase, which is appropriate according to the aforementioned assumptions. The authors recommend, therefore, that Eq. 32 should be employed for gas permeation studies, in which no liquid phase is present. For pervaporation experiments, however, Eq. 38 can adequately describe the phenomena.

$$J_i = \frac{Q_i^L}{\lambda_M} \cdot \left(X_{i,F} - \frac{P_{i,P}}{H_i}\right) \tag{38}$$

2.3. Membrane Synthesis

As previously stated in this chapter, usually the use of non-porous membranes is employed for pervaporation assays, in which the permeation is governed by mass diffusion, unlike other membrane separation processes, in which the porous membranes require pressure difference between the phases. A summary of the membranes that can be employed for pervaporation use is shown in Figure 2.5. It is clear that two broad categories can be outlined for the membranes: symmetric and asymmetric. The main difference, as one can expect, is due to the membranes structure and morphology. Symmetric membranes are homogeneous in its structure, that is, no change whatsoever in the membranes porosity or material is present throughout its thickness. Asymmetric membranes, on the other hand, consist of a thin non-porous membrane placed on a porous sheet, which acts as a support layer, providing adequate mechanical properties such as elasticity, tensile strength and elongation at break. At this point, it is important to point out that the non-porous membrane layer also presents interstitial vacancies on its structure. However, opposed to the porous membranes, those are present only in a molecular scale, which does not affect the mass transport throughout the membrane, but are of great importance due to the promotion of membrane swelling, which is essential for an efficient permeation (Gimenes et al., 2007).

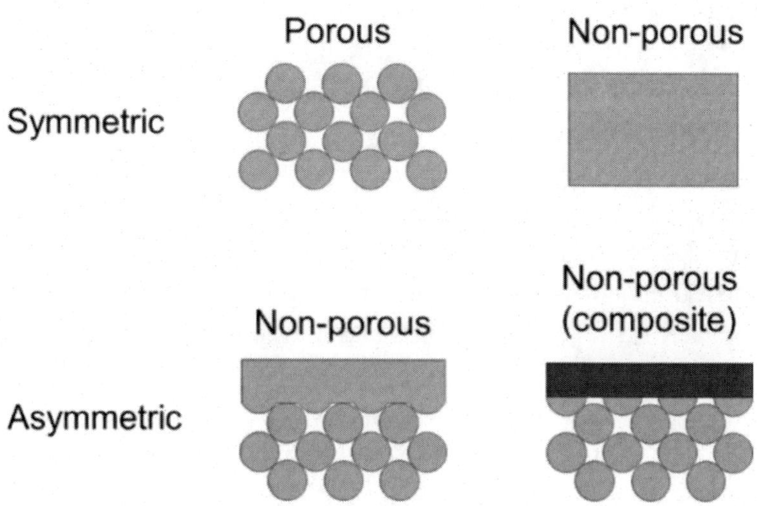

Figure 2.5. Membrane types applicable to pervaporation experiments.

Figure 2.5 also indicates that the membranes employed in pervaporation can be classified according to the materials employed in the synthesis procedure. When two different materials are employed for the dense and porous layers, the asymmetric membrane synthesized is classified as composite non-porous membranes. Those are the most frequently membranes employed in pervaporation experiments, since they present the advantage of the layers presenting different properties.

It is comprehensible that the membranes composition, structure as well as morphology are of much influence to the membranes selectivity, since the interaction of the components present in the feed solution with the membrane are dependent on its properties. Therefore, the synthesis procedures have received increased attention over the years, since the optimization of this process greatly contributes to the overall permeation selectivity.

Among the optimization of the membranes synthesis, the choosing of the adequate material is of great importance and should take into consideration the characteristics of the feed solution as well as the process demands. Long-chain polymers like polyvinyl alcohol (PVA) and polydimethylsiloxane (PDMS) are commonly employed, showing adequate selectivity for wide pervaporation applications. This type of molecules, due to an increased intermolecular cohesion, usually promotes desirable mechanical properties such as elasticity, tensile strength and elongation at break. Those distinguished properties emerge from specific knowledge of chemistry of polymers, since the induced chemical crosslinking alters the polymer glass transition temperature among other properties. A more detailed description of the chemistry involved in the membrane synthesis, despite its importance in membrane characterization, is beyond the scope of this chapter. However, the reader is encouraged to seek specified literature about this topic.

Among the several synthetization procedures employed for membrane production, the casting method deserves some attention in this chapter for being a low-cost, simple and effective procedure. A previously prepared film forming solution, consisting mainly of a polymer blend, crosslinking agent and a solvent (usually water) is casted on a dish, followed by the removal of the solvent by low-temperature evaporation, desirably employing a vacuum-drying system. A previously determined mass of the film forming solution yields a specific dense layer thickness once the solvent is evaporated. Composite asymmetric dense membranes are also synthesized by this method. This procedure is frequently employed for pervaporation membranes production since a non-porous membrane is obtained at the end.

As a final remark, due to the growing environmental concern, the employment of biopolymers has received increased attention over the last years, with membranes manufactured form industrial byproducts such as silk sericin and polysaccharides like chitosan and sodium alginate being more frequent. This alternate production presents the advantage of dispensing nonrenewable resources, whose recently increased pricing has offered economical difficulties. In addition, industrial byproducts can be a great local environmental concern, since the uneven distribution of natural resources promotes the production of large amounts of wastes and byproducts in a specific area. Therefore, the alternative application of these resources in the biopolymer industry can contribute to the reduction of organic material being, therefore, much environmentally friendly.

2.4. Membrane Selectivity in Pervaporation Process

The efficiency of the pervaporation process, also known as the membranes selectivity, represents the extent of the permeation of the desired component. Historically, the separation factor, α, has been employed to this aim. It relates the compositions of the permeate and feed solution as given by Eq. 39, in which the subscripts i and j correspond, respectively, to the more permeable component present in the solution and the less permeable one. It is clear therefore, that the separation is inefficient when the separation factor approaches the unity, whereas the larger the separation factor obtained, the more efficient is the process.

$$\alpha = \frac{y_i/y_j}{x_i/x_j} \tag{39}$$

For multicomponent solutions, however, the identification, as well as quantification, of the said i and j components can be difficult. Under these circumstances, the membranes selectivity is more easily described by the enrichment factor, as given by Eq. 40, which compares the concentration of the i component in the permeate and feed. The enrichment factor, therefore, represents the increase of the components concentration in the process and is normally calculated for all the components of interest. The separation factor, on the other hand, being more applicable for bicomponent solutions, represents the overall permeation efficiency, being a global parameter of the membranes selectivity.

$$\beta_i = \frac{y_i}{x_i} \tag{40}$$

However, as the previous sections indicate, the selectivity of the membrane depends on a large number of variables. Intrinsic properties of the membrane such as composition and synthesis method are major factors that influence the membranes selectivity. In addition, operational factors like temperature, feed composition and vacuum intensity are also of much influence. Therefore, both the separation factor and enrichment factor are dependent on all these aspects. Changing one of the operational parameters, and the selectivity will also change. This offers some difficulties, since a thorough comparison between different pervaporation membranes becomes difficult, since all the operational parameters have to be the same. It should be more interesting, then, to compare the selectivity of the process by a parameter that is inherent to the membrane itself, being less dependent of operating conditions. This parameter, as proposed by Wijmans et al. (2009) is the permeability, which represents a preferred way for reporting pervaporation data, as Eq. 41 suggests. Therefore, the authors recommend the use of the aforementioned separation factor, since the permeability (Q) or, more precisely, the difference in permeability of the components represents the overall process efficiency.

$$\alpha_i = \frac{Q_i}{Q_j} \tag{41}$$

3. Applications of Pervaporation

Pervaporation when compared to conventional processes such as distillation, liquid-liquid extraction and adsorption, stands out as an interesting separation technology. As a membrane separation process, it can be considered as a clean technology and offers energy saving potential due to reduced temperature and pressure requirements. Pervaporation has been employed in the dehydration of alcohols and organic solvents, removal of organic compounds from solutions and separation of anhydrous organic mixtures. Due to the use of high selective membranes, pervaporation can remove small amounts of organic compounds from contaminated water. Other potential uses of pervaporation are the recovery of aromas, the de-alcoholization of beer in the food industry and the recovery of fermentation products. Table 1 shows some applications of pervaporation, as well as the membrane and the feed solution used in each study.

Table 1. Some applications of pervaporation

Application	Membrane	Feed mixture	Separation factor	Permeate flux (g/m^2h)	Reference
Dehydration	NH_2-UiO-66/PEI	Acetic acid/ water (95/5)	356	212	Wang et al., 2017
	Silicalite 1-PDMS	Ethanol/water (5/95)	34.3	176	Zhuang et al., 2016
	PDMSM/PTMSP	Butanol 2%	128	120	Borisov et al., 2014
	Silica/PTMSP	Butanol 5%	104	9500	
	Polybenzoxazole (PBO)	Ethanol/water (85/15)	85	82	Ong et al., 2012
Organics separation	PDMS/PEI	Benzene/cyclohexane	13.2	218	Chen et., 2009
	SA25C75	Benzene/cyclohexane	57.9	2233	Kuila and Ray, 2014
	Polybenzoxazole	Toluene/ heptane (40/60)	3.7	70	Ribeiro et al., 2012

The efficiency of pervaporation is determined by the properties of the substance to be recovered and the membrane, depending also on the operating variables such as feed temperature, permeate pressure and feed concentration. Understanding the effects of these operational variables is of fundamental importance in order to obtain an effective separation. Although the examples of pervaporation processes shown in Table 1 have not occurred under the same conditions, the variety of membranes and new materials is evident. New materials and membrane configurations have been explored in the search for more efficient processes.

3.1. Dehydration of Organic Solvents

The hydrophilic membranes were the first one to achieve industrial applications and were used for the organic solvent dehydration. Commercially hydrophilic membranes earlier made of polymeric materials such as natural polymers like cellulose and its derivatives were subsequently substituted by synthetic materials. Polyvinyl alcohol (PVA), poly (acrylic acid) (PAA), polyacrylonitrile (PAN), and nylon 6 have been investigated mainly on the ethanol dehydration (Jyoti et al., 2015; Shao and Huang, 2007). Many other hydrophilic materials were studied in the solvent dehydration, including sodium alginate (SA), chitosan, polyethyleneimine (PEI) and synthetic zeolites.

Sodium alginate is a polysaccharide obtained from seaweeds, and it shows excellent affinity for water. Song et al. (2018) prepared hybrid membranes by incorporating hydrophobic 2HMoS2 nanosheets within hydrophilic SA polymer matrix. The ethanol dehydration performance was 54% higher when compared to a SA/PAN pure membrane. The separation factor was 1229 (85% greater than that of SA/PAN pure membrane).

Ong et al. (2016) reported the potential of aromatic polymers such as polybenzoxazole (PBO), polybenzoxazinone (PBOZ) and polybenzimidazole (PBI) in solvent dehydration. They have excellent chemical and thermal resistance. Both PBO and PBOZ membranes were found to be applicable in the dehydration of alcohols. PBO exhibited a stable performance in dehydration of isopropanol and n-butanol at 80 °C throughout the experimental period of 250 h (Ong et al. 2012).

Zeolite membranes have also been used for dehydration assays due to their high separation performance and good thermal/chemical stability. Comercial NaA zeolite membranes and other zeolite membranes with higher Si/Al ratio showed high selectivity and permeation flux when compared with polymeric membranes. Hydrophilic zeolite membranes have strong adsorption to water, showing very high water separation selectivity.

Zhang et al. (2018) prepared and tested an acid resistant DD3R zeolite pervaporation membrane for the dehydration of acetic acid. They achieved a water permeation flux of 0.58 kg m^{-2} h^{-1} as well as a separation factor of 800 at 95 °C, confirming the excellent acid-resistance of the membrane. The dehydration of acetic acid is a critical process in the industrial production of acid. The membranes used for acetic acid dehydration include polymeric and inorganic materials. PVA is the most commonly polymeric material used for acetic acid dehydration. It conferes chemical stability, hydrofilicity and good tensile strenght to the membrane.

It has been observed in recent years that research on pervaporation membranes for solvent dehydration seeks new chemically and thermally stable materials to dehydrate aggressive solvents at elevated temperatures.

3.2. Recovery of Organics

Unlike dehydration, the removal of dilute organic compounds from aqueous solutions by pervaporation requires the preferential permeation of organic compounds through the membrane. For this, the membrane has to be made of organophilic or hydrophobic materials to increase the affinity of the organic compounds towards it. In addition, the application of pervaporation in the recovery of aromatic compounds has the advantage of operating at low temperatures, which avoids the degradation of thermally unstable but high value compounds (Ong et al, 2016).

Polydimethylsiloxane (PDMS) is the most widely studied hydrophobic material used to recover organics from aqueos solutions. It shows high affinity and low transport resistance for organics, and is also very stable in the streams. Other hydrophobic materials such as styrene based polymers, polyvinylidene fluoride (PVDF), polyether-block-polyamides (PEBA), etc. were also reported for this application.

Although hydrophobic pervaporation is very useful in recovering alcohol and other organic compounds, there are still few available hydrophobic materials that have attractive performance. To improve the interaction between organic compounds with the polymer matrix, novel mixed matrix membranes (MMMs) has been studied. The mostly common inorganic fillers are ZSM-5 zeolite, silicalite and other hydrophobic particles. They had capacity to improve the performance of silicone rubber membranes for alcohol recovery (Vane et al., 2008; Gu et al., 2009; Huang et al., 2009).

Mixed matrix membranes (MMMs) prepared using polydimethylsiloxane (PDMS) are also tested for the separation of ethanol/water mixtures via pervaporation. Compared to unfilled PDMS membranes, MMMs improves the flux and the separation factor. Khan et al. (2018) used zeolitic imidazole frameworks (ZIF-67) as inorganic filler particle to prepare MMMs for a variety of membrane applications. The pervaporation performance of the MMMs was evaluated for 6 wt% aqueous ethanol solution. Compared to pristine PDMS membranes (0% filler load), membranes filled with 20% ZIF-67 showed a triple increase in flux and twice in separation factor at 40 °C.

3.2.1. Recovery of Aroma Compounds

Aroma and aroma compounds are examples of non-alcoholic organic compounds which can be recovered by pervaporation. Many studies with different types of membranes have been reported and tested on the recovery of non-alcoholic organic compounds from aqueous solutions. Isopropyl acetate was pervaporated from an aqueous solution (0.39 wt%) using ZSM-5 filled hydroxyl terminated polybutadiene (HTPB)-based polyurethaneurea (PU) membranes (Zhang et al., 2012). Pervaporation process as a single step of downstream operation can recover and concentrate compounds from fermentation broths. A commercial polymeric organophilic membrane with a PDMS separation film, PERVAP™ 4060 was tested to concentrate the isoamyl acetate produced by fermentation of sugarcane molasse. The analysis of the enrichment factors indicated a decrease in the selectivity of the membrane to the ester isoamyl acetate. The total flow also decreased after 3 h of pervaporation, remaining constant at 800g $m^{-2}h^{-1}$. The obtained product became colorless and had a characteristic banana flavor (Rossi et al, 2017).

The pervaporation process can be used to recover aroma compounds lost during industrial processing of juices and beers. Olmo et al. (2014) used a comercial PDMS membrane with hidrophobic/organophilic characteristics to recover aromas of beer and added to a low-alcohol beer and an alcohol-free beer. Three flavor constituents of beer (isobutyl alcohol, ethyl acetate and isoamyl acetate) were selectively recovered through the membrane and incorporated on non-alcohol beers to improve their sensory quality. Ethyl acetate had the higher enrichment factor (around 40) after 2h of pervaporation. The non-alcoholic beers were sensory evaluated. The results showed that 90% of tasters preferred enriched low-alcohol beer instead of low-alcohol beer and 80% of tasters preferred enriched alcohol-free beer instead of alcohol-free beer. This is an example of the applicability of pervaporation to selectively concentrating compounds of interest. In this case, the ethanol was discarded by the membrane and the aroma compounds recovered.

Pervaporation can also been utilized in industrial food processing. As an example, shellfish flavor is a high value food product. Some aroma compounds identified in the brown crab boiling juice were concentrated from a model dilute aqueous solution. This process with pervaporation membrane PERVAP 4060 recovery 1-octen-3-ol (86%), 1-penten-3-ol (81%), 3-methylbutanal (24%), benzaldehyde (94%), 2,3-pentanedione (21%), hexanal (44%) and ethyl acetate (54%) (Martinez, Sanz and Beltran, 2012).

POMS and PDMS membranes were tested for pervaporation experiments of binary and ternary model solutions of pomegranate aroma compounds. The effect of the membranes thickness on the permeation flux and aroma enrichment factors was investigated. The results showed that the permeation flux through both membrane types decreased markedly with increasing the membrane thickness, while the aroma compound enrichment factors increased. In addition, the composition of the feed solution is of great influence to the pervaporation process. It was observed that the permeation of some aroma compounds was affected by presence of other aromatic molecules (Raisi and Aroujalian, 2011).

In the production of instant soluble coffee, the most important aromas are lost during the process, endorsing the applicability of pervaporation as an alternative processing that can be adopted to increase the aromatic perception of instant coffee. To this aim, pervaporation experiments with PDMS membranes have been succesfully carried out. Weschenfelder et al. (2015) were able to depic the key aromas of soluble coffee as being the following compounds: 2,3-Butanedione (sweet, buttery, creamy and milky), 2,3-pentanedione (sweet buttery, creamy, slightly toasted dairy), 3-methyl-butanal (cheesy, sweaty and fruity), benzaldehyde (sweet, oily, almond and cherry), acetaldehyde (purgent), furfural(vegetable, burnt astringent nuance), 2,5-dimethylpirazine(nutty-like), 5-methyl furfural (sweet, spicy, warm odor with a sweet, caramel-like). Permeances through the membrane (PDMS with a polyethylene terephthalate (PET) support layer material) was estimated for the coffee aroma compounds mentioned above. Results showed that pervaporation is a promising alternative to concentrate aroma compounds from soluble coffee.

3.3. Separation of Organic Compounds

Pervaporation also found apllication as a method to separate mixtures of organic compounds. The industrial needs to separate mixtures of benzene/ ciclohexane, toluene/heptane, p-xylene/o-xylene, and others with similar properties, showed some promising pervaporation attempts (Shao and Huang, 2007). The most explored materials for separating aromatic/aliphatic compounds are organic polymers.

The separation of methanol and butyl acetate from the reaction media increases the productivity when utilizing a hybrid configuration composed of reaction pervaporation working under the most appropriate operation conditions to obtain competitive results against distillation. Luis et al. (2013) tested different membranes to separate butyl accetate from methanol. PolyAn membranes and the Pervap 2255–50 membrane shown best performance when the concentration of the feed solution is rich in butyl acetate. In particular, the membranes Poly OL M1 and M2 achieve the highest permeance for butyl acetate. On the other hand, the membrane Pervap 1201 is the best choice when an intermediate or low concentration of butyl acetate is present in the mixture.

3.4. Integrated Process

Pervaporation is extensively used in chemical industries as a separation process. However, the integration with reactional processes have also been carried out over the years. Such hybrid systems, normally designated as membrane reactors, are commonly integrated with pervaporation membranes, in which the simultaneous reaction and separation processes provides several operational advantages. Usually these improvements are related to the increased separation efficiency due to the chemical reactions present or, more frequently, to the enhanced reactants conversion and reaction extent due to the simultaneous removal of reactants or products from the system (e.g., overcoming equilibrium constraints, enhancing the reaction rates, increasing selectivity). Therefore, such improvements allow an optimization of the integrated equipments design when comparing to the individual systems, since the higher efficiency induces lower energy demands and enable a more compact design. In an in-situ pervaporation reactor both reaction and separation are performed in the same unit. Recently, the applicability of integrated membrane reactor has been studied for esterification reactions. In these systems, which are limited by thermodynamical equilibrium, water is produced as a byproduct, reducing the forward reaction rate and the formation of the product of interest. The difficulty can be overcome by the employment of integrated pervaporation, which can drive the equilibrium reaction enhancing the conversion due to the continuously removal of one of the products (water or ester) (Chandane et al., 2017).

In an in-situ type of pervaporation reactor, a polymeric hydrophilic PVA–PES membrane assisted the esterification of propionic acid and isobutyl alcohol, producing isobutyl propionate (characteristic compound of apricot, apple and cherry flavor). This process presented as a more effcient alternative, yielding a larger conversion than the batch process of esterification. The hydrophilic membrane showed high selectivity to the removal of water in the propionic acid, isobutyl alcohol, isobutyl propionate and water mixture (Chandane et al., 2017).

In another study, the esterification of propionic acid with ethanol was successfully conducted in a pervaporation membrane reactor. A polyvinyl alcohol membrane was used to simultaneously remove the water from the reaction medium. Conversion results of a membrane reactor were compared to results from a batch reactor. The highest conversion was obtained in the membrane reactor (Nigiz, 2018). Kulkarni et al. (2010) selectively recovered limonene from orange peels free of solvent. The authors tested an integrated process of extraction with polypropylene glycol (PPG) followed by organophilic pervaporation using a polyoctylmethylsiloxane on polyetherimide (POMS-PEI) membrane. Extracts in polyethylene glycol 300 as feed solutions, results in a factor of enrichment for limonene of 237.8 and a partial flux of 0.051 $mol.m^{-2}. h^{-1}$.

The organophilic pervaporation was also utilized to remove a solvent used in the immobilization of 2-phenylethanol, a flavor molecule with a rose-like smell produced by fermentation. The use of a pervaporation unit containing a polyoctyl methylsiloxane membrane increased the 2-PE production by nearly 100% (Hua and Xu, 2011; Carlquist et al., 2015).

Conclusion

Despite the numerous studies and stablished processes revealing the potential of membrane separation processes, the dehydration of organic solvents remains the main industrial application of pervaporation. The recovery and purification of solvents are still little explored in an industrial scale. This can be explained by the limited availability of commercial membranes that are that are chemically and mechanically resistant and have high performance. This area still lacks on studies that enable the development of new membrane materials and explain the mechanism of transport across the membranes. The need to develop cheaper and more competitive technologies led to the idea of "process intensification", with the design of compact catalytic reactors, new process projects or reorganization of current process projects. This includes integrated operations that combine production and separation of products or by-products. This configuration can increase the process productivity and selectivity, decrease energy consumption, and result in high efficiency systems with environmentally friendly engineering features.

REFERENCES

Baker, R.W., Wijmans, J.G., (1995). The solution-diffusion model: a review. *J. Memb. Sci.* 107, 1-21.

Binning, R., Lee, R., Jenning, J., Martin, E., (1961). Separation of liquid mixtures by permeation. *Ind. End. Chem. Res.* 53, 45-50.

Borisov, I.L., Malakhov, O.A., Khotimsky, V.S., Litvinova, E.G., Finkelshtein, E.S., Ushakov, N.V., Volko, V.V., (2014). Novel PTMSP-based membranes containing elastomeric fillers: Enhanced 1-butanol/ water pervaporation selectivity and permeability. *J. Memb.Sci.* 466, 322-330.

Carlquist, M., Gibson, B., Yuceer, Y.K., Paraskevopoulou, A., Sandell, M., Angelov, A.I., Gotcheva, V., Angelov, A.D., Etschmann, M., Billerbeck, G.M.,Lidén, G., (2015). Process engineering for bioflavour production with metabolically active yeasts – a mini-review. *Yeast* 32,123-143.

Chandane, V.S., Rathod, A.P., Wasewar, K.L., (2016). Enhancement of esterification conversion using Pervaporation membrane reactor. *Res.-Effic. Technol.* 2, 47- 52.

Chandane, V.S., Rathod, A.P., Wasewar, K.L., (2017). Coupling of in-situ pervaporation for the enhanced esterification of propionic acid with isobutyl alcohol over cenosphere based catalyst. *Chem. Eng. Proc.-Process Intensification* 119, 16-24.

Chen, J., Li, J., Lin, Y., Chen, C., (2009). Pervaporation performance of polydimethylsiloxane membranes for separation of benzene/ cyclohexane mixtures. *J. Appl. Polym. Sci.*, 112, 2425-2433.

Chen, X., Li, W., Shao, Z., Zhong, W., Yu, T., (1999). Separation of Alcohol-Water Mixture by Pervaporation Through a Novel Natural Polymer Blend Membrane Chitosan/Silk Fibroin Blend Membrane. *J. App. Pol. Sci.* 73, 975-980.

Feng, X., Huang, R.Y.M.; (1997). Chen, X., Li, W., Shao, Z., Zhong, W., Yu, T., (1999). Liquid separation by membrane pervaporation: a review. *Ind. End. Chem. Res.* 25, 1048-1066.

Gimenes, M.L., Liu, L., Feng, X., (2007). Sericin/poly(vinyl alcohol) blend membranes for pervaporation separation of ethanol/water mixtures. *J. Memb. Sci.* 295, 71-79.

Gnansounou, E., Dauriat, A., (2005). Gimenes, M.L., Liu, L., Feng, X., (2007). Ethanol fuel from biomass: a review. *J. Sci. Ind. Res.* 64, 809-821.

Gu, J., Shi, X., Bai, Y.X., Zhang, H.M., Zhang, L., Huang, H., (2009).Silicalite filled polyether block-amides membranes for recovering ethanol from aqueous solution by pervaporation. *Chem. Eng. Technol.* 32, 155-160.

Heissler, E.G., Hunter, A. S., Sciliano, J., Treadway, R.M., (1956). Solute and temperature effects in the pervaporation of aqueous alcoholic solutions. *Sci.* 124, 77-79.

Hua, D., Xu, P., (2011). Recent advances in biotechnological production of 2-phenylethanol. *Biotechnol. Advances* 29, 654-660.

Huang, Y.W., Zhang, P., Fu, J.W., Zhou, Y.B., Huang, X.B., Tang, X.Z., (2009). Pervaporation of ethanol aqueous solution by polydimethyl-siloxane/polyphosphazene nanotube nanocomposite membranes. *J. Membr. Sci.* 339, 85-92.

Jyoti, G., Keshav, A., Anandkumar, J., (2015). Review on Pervaporation: Theory, Membrane Performance and Application to Intensification of Esterification Reaction. *J. Eng.* 2015, 24 p. Available from: http://dx.doi.org/10.1155/2015/927068.

Khan, A., Ali, M., Ilyas, A., Naik, P., Vankelecom, I.F.J., Gilani, M.A., Sajjad, Z., Khan, A.L., (2018). ZIF-67 filled PDMS mixed matrix membranes for recovery of ethanol via pervaporation. *Sep. Purif. Technol.* 206, 50-58.

Kober, P.A., (1917). Pervaporation, perstillation and percrystalization. *J. American Chem. Soc.* 39, 944-948.

Koretsky, M.D., (2007). *Engineering and Chemical*, John Wiley & Sons, New Jersey.

Kuila, S.B., Ray, S.K., (2014). Dehydration of dioxane by pervaporation using filled blend membranes of polyvinyl alcohol and sodium alginate. *Carbohyd. Polym.* 101, 1154-1165.

Kulkarni, P.S., Brazinha, C., Afonso, C.A.M. Crespo, J.G., (2010). Selective extraction of natural products with benign solvents and recovery by organophilic pervaporation: fractionation of D-limonene from orange peels. *Green Chem.*, 12, 1990-1994.

Lue, S.J., Wang, S.F., Wang, L.D., Chen, W.W., Du, K., Wu, S.Y., (2008). Diffusion of multicomponent vapor in a poly(dimethyl siloxane) membrane. *Des.* 233, 277-285.

Luis, P., Degrève, J., Bruggen, B.V., (2013). Separation of methanol–n-butyl acetate mixtures by pervaporation: Potential of 10 commercial membranes. *J Memb. Sci.* 429, 1–12.

Martínez, R., Sanz, M.T., Beltrán, S., (2013). Concentration by pervaporation of brown crab volatile compounds from dilute model solutions: Evaluation of PDMS membrane. *J. Memb. Sci.* 428, 371-379.

Néel, J., (1995). Pervaporation. In Noble, R.W. & Stern, S.A. (Eds.), *Membrane Separations Technology: Principles and Applications*. New York: Elsevier.

Nigiz, F.U., Dogan, H., Himioglu, N.D., (2012). Pervaporation of ethanol/water mixtures using clinoptiolite and 4A filled sodium alginate membranes. *Des.* 300, 24-31.

Nigiz, F.U., Himioglu, N.D., (2012). Pervaporation of ethanol/water mixtures by zeolite filled sodium alginate membrane. *Des. Water Treat.* 51, 637-643.

Nigiz, F.U., (2018). A comparative study on the synthesis of ethyl propionate in a pervaporation membrane reactor. *Chem. Eng. Proc.* 128, 173-179.

Olmo, A., Blanco, C.A., Palacio, L., Pradanos, P., Hernandez, A., (2014). Pervaporation methodology for improving alcohol-free beer quality through aroma recovery. *J. Food Eng.* 133, 1-8.

Ong, Y.K., Shi, G.M., Le, N.L., Tang, Y.P., Zuo, J., Nunes, S.P., Chung, T.-.S., (2016). Recent membrane development for pervaporation processes. *Progr. Polymer Sci.* 57, 1-31.

Ong, Y.K., Wang, H., Chung, T.S., (2012). A prospective study on the application of thermally rearranged acetate-containing polyimide membranes in dehydration of biofuels via pervaporation. *Chem. Eng. Sci.* 79, 41-53.

Raisi, A., Aroujaliana, A., Kaghazchi, T., (2008). Multicomponent pervaporation process for volatile aroma compounds recovery from pomegranate juice. *J. Memb. Sci.* 322, 339-348.

Ribeiro, C.P., Freeman, B.D., Kalika, D.S., Kalakkunnath, S., (2012). Aromatic polyimide and polybenzoxazole membranes for the fractionation of aromatic/aliphatic hydrocarbons by pervaporation. *J. Membr. Sci.*, 390-391, 182-193.

Rossi, S.C., Medeiros, A.B.P., Weschenfelder, T.A., Scheer, A.P., Soccol, C.R., (2017). Use of pervaporation process for the recovery of aroma compounds produced by P.fermentans in sugarcane molasses. *Biop. Biosyst. Eng.* 40, 959-967.

Sandler S.I., (1999). *Chemical and Engineering Thermodynamics* (3rd edition). New Jersey, John Wiley & Sons.

Shao, P., Huang, R.Y.M., (2007). Polymeric membrane pervaporation. *J. Membr. Sci.* 287, 162-179.

Song, Y., Jiang, Z., Gao, B. Wang, H., Wang, M., He, Z., Cao, X., Pan, F., (2018). Embedding hydrophobic MoS2 nanosheets within hydrophilic sodium alginate membrane for enhanced ethanol dehydration.*Chem. Eng. Science* 185, 231-242.

Vane, L.M., Namboodiri, V.V., Bowen, T.C.,(2008). Hydrophobic zeolite-silicone rubber mixed matrix membranes for ethanol-water separation:

effect of zeolite and silicone component selection on pervaporation performance. *J. Membr. Sci.* 308,230-234.

Wang, N., Zhang, G., Wang, L., Li, J., An, Q., Ji, S., (2017). Pervaporation dehydration of acetic acid using NH_2- UiO-66/PEI mixed matrix membranes. *Sep. Purif. Technol.* 186, 20-27.

Weschenfelder, T.A., Lantin, P., Viegas, M.C., Castilhos, F., Scheer, A.P., (2015). Concentration of aroma compounds from an industrial solution of soluble coffee by pervaporation process. *J. Food Eng.* 159, 57- 65.

Wijmans, J.G., Baker, R.W., (2006). The solution-diffusion model: A unified approach to membrane permeation. In_ Yampolskii, Y., Pinnau, I., Freeman, B., (eds.), *Materials Science of Membranes for Gas and Vapor Separations,* John Wiley & Sons, New Jersey.

Wijmans, J.G., Baker, R.W., Huang, Y., (2009). Permeability, permeance and selectivity: A preferred way of reporting pervaporation performance data. *J. Memb. Sci.* 348, 346-352.

Zhang, Y., Chen, S., Shi, R., Du, P., Qiu, X., Gu, X., (2018). Pervaporation dehydration of acetic acid through hollow fiber supported DD3R zeolite membrane. *Sep. Purif. Technol.* 204, 234-242.

Zhang, W.-d., Sun, W., Yang, J., Ren, Z.-q., (2010). The study on pervaporation behaviors of dilute organic solution through PDMS/PTFE composite membrane. *Appl. Biochem. Biotechnol.* 160, 156-167.

Zhang, C.F., Yang, L., Bai, Y.X., Gu, J., Sun, Y.P., (2012). ZSM-5 filled polyurethaneurea membranes for pervaporation separation isopropyl acetate from aqueous solution. *Sep. Purif. Technol.*85, 8-16.

Zhuang, X., Chen, X., Su, Y., Luo, J., Feng, S., Zhou, H., Wan, Y., (2016). Surface modification of silicalite-1 with alkoxysilanes to improve the performance of PDMS/silicalite-1 pervaporation membranes: Preparation, characterization and modeling. *J. Memb. Sci.* 499, 386-395.

BIOGRAPHICAL SKETCHES

Adriane Bianchi Pedroni Medeiros

Affiliation: Department of Bioprocess Engineering and Biotechnology, Federal University of Paraná. Curitiba, Paraná, Brazil

Education: PhD in Bioprocess Engineering and Biotechnology (Federal University of Paraná – 2003)

Honors: CNPq Research Productivity Scholarship

Articles from the Last 3 Years:

1. Sydney, E.B.; Novak, A.C.; Rosa, D.; Medeiros, A.B.P.; Brar, S.K.; Larroche, C.; Soccol, C.R. Screening and bioprospecting of anaerobic consortia for biohydrogen and volatile fatty acid production in a vinasse based medium through dark fermentation. *Process Biochemistry,* v. x, p. 1-10, 2018.
2. Rossi, S.C.; Medeiros, A.B.P.; Weschenfelder, T.A.; Scheer, A.P.; Soccol, C.R. Use of pervaporation process for the recovery of aroma compounds produced by P. fermentans in sugarcane molasses. *Bioprocess and Biosystems Engineering,* v. 40, p. 959-967, 2017.
3. Matos, M.E; Medeiros, A.B.P.; Pereira, Gilberto Vinícius De Melo; Soccol, Vanete Thomaz; Soccol, Carlos R. Production and Characterization of a Distilled Alcoholic Beverage Obtained by Fermentation of Banana Waste (Musa cavendishii) from Selected Yeast. *Fermentation,* v. 3, p. 62, 2017.
4. Bier, M.C.J.; Medeiros, A.B. P.; De Oliveira, J.S.; Côcco, L.C.; Costa, J.; Carvalho, J.C.; Soccol, C.R. Liquefied gas extraction: A new method for the recovery of terpenoids from agroindustrial and forest wastes. *The Journal of Supercritical Fluids,* v. 110, p. 97-102, 2016.
5. Spier, M.R.; Siepmann, F.B.; Staack, L.; Souza, P.Z.; Kumar, V.; Medeiros, A.B. P.; Soccol, C.R. Impact of microbial growth inhibition

and proteolytic activity on the stability of a new formulation containing a phytate-degrading enzyme obtained from mushroom. *Preparative Biochemistry & Biotechnology,* doi: 10826068.2015.1, 2016.
6. Fiorda, F.A.; Pereira, G.V.M.; Thomaz-Soccol, V.; Medeiros, A.B.P.; Rakshit, S.; Soccol, C.R. Development of kefir-based probiotic beverages with DNA protection and antioxidant activities using soybean hydrolyzed extract, colostrum and honey. *Lebensmittel-Wissenschaft + Technologie/Food Science + Technology,* v. 68, p. 690-697, 2016.
7. Pereira, G.V. M.; Carvalho Neto, D.P.; Medeiros, A.B.P.; Soccol, V.T.; Neto, E.; Woiciechowski, A.L.; SOCCOL, C.R. Potential of lactic acid bacteria to improve the fermentation and quality of coffee during on-farm processing. *International Journal of Food Science & Technology (Print),* v. 51, p. 1689-1695, 2016.
8. Soccol, C.R.; Dalmas Neto, C.J.; Soccol, V.T.; Sydney, E.B.; Costa, E. F.S.; Medeiros, A.B.P.; Vandenberghe, L.P.S. Pilot scale biodiesel production from Rhodosporidium toruloides DEBB 5533 microbial oil using sugarcane juice: Performance in diesel engine and preliminary economic study. *Bioresource Technology,* v. 223, p. 259-268, 2016.
9. Bier, M.C.J.; Medeiros, A.B. P.; Soccol, C.R. Biotransformation of limonene by an endophytic fungus using synthetic and orange residue-based media. *Fungal Biology,* v. x, p. 1-8, 2016.

Joel Robert Karp

Affiliation: Graduate Program in Mechanical and Materials Engineering, Federal University of Technology – Paraná. Dep. Alencar Furtado St., Curitiba, Paraná, Brazil

Education: MSc. in Food Engineering (Federal University of Paraná – 2017). Bachelor in Chemical Engineering (Federal University of Paraná – 2015)

Publications from the Last 3 Years:

KARP, J. R.; HAMERSKI, F.; SILVA, V. R. Supported Silk Fibroin/Poly(vinyl alcohol) Membrane Blends: Structure, Properties, and Ethanol Dehydration by Pervaporation. *Polymer Engineering and Science*, DOI 10.1002/pen.24796

Vitor Renan da Silva

Vitor Renan da Silva graduated in Chemical Engineering in 2006 from Federal University of Parana (UFPR). He obtained master degree in Food Technology from Federal University of Paraná in 2009, and Dr. Eng., Food Engineering from Federal University of Paraná in 2013.

Vitor worked in Environmental Department of Curitiba, Brazil, between 2011 and 2014, and since 2014 has been working as adjunct professor in chemical engineering department of UFPR. His main research areas are industrial separation process including adsorption, microfiltration, ultra-filtration and pervaporation.

Moisés A. Marcelino Neto

Affiliation: Mechanical Engineering Department, Federal University of Technology - Paraná, Dep. Alencar Furtado St., Curitiba, Paraná, Brazil

Education: Federal University of Santa Catarina, Florianópolis, Santa Catarina, Brazil.

PhD Degree in Mechanical Engineering, 2012 – 2013, Particular area of study: Thermodynamics.

DSc Degree in Mechanical Engineering, 2007 – 2011. Particular area of study: Thermodynamics and Mass Transfer.

Master's Degree in Mechanical Engineering, 2004 – 2006. Particular area of study: Fluid Phase Equilibria.

Federal University of Ceará, Fortaleza, Ceará, Brazil:

Degree in Mechanical Engineering, 1999 – 2003

Publications from the Last 3 Years:

1. Sirino, T.H.; Bertoldi, D.; Marcelino Neto, M.A.; Sum, A.K.; Morales, R.E. M. *Multiphase Flash Calculations for Gas Hydrates Systems.* Fluid Phase Equilibria (In Press), 2018.
2. Rosas, L.M. M.; Bassani, C.L.; Alves, R.F.; Schneider, F.A.; Marcelino Neto, M.A.; Morales, R.E. M.; Sum, A.K. Measurements of Horizontal Three-Phase Solid-Liquid-Gas Slug Flow: Influence of Hydrate-Like Particles on Hydrodynamics. *AICHE Journal,* v. 64, p. 2864-2880, 2018.
3. Antunes, C.M.M.O.; Kakitani, C.; Marcelino Neto, M.A.; Sum, A.K.; Morales, R.E. M. An Examination of the Prediction of Hydrate Formation Conditions in the Presence of Thermodynamic Inhibitors. *Brazilian Journal of Chemical Engineering,* v. 35, p. 265-274, 2018.
4. Andrade, D.E. V.; Marcelino Neto, M.A.; Negrao, C.O. The importance of supersaturation on determining the solid-liquid equilibrium temperature of waxy oils. *Fuel,* v. 206, p. 516-523, 2017.
5. Guembaroski, A.Z.; Marcelino Neto, M.A.; Bertoldi, D.; Morales, R.E. M.; Sum, A.K. Phase Behavior of Carbon Dioxide Hydrates: A Comparison of Inhibition Between Sodium Chloride and Ethanol. *Journal of Chemical and Engineering Data,* v. 62, p. 3445-3451, 2017.
6. Ferrari, P.F.; Guembaroski, A.Z.; Marcelino Neto, M.A.; Morales, R.E.M.; Sum, A.K. Experimental measurements and modelling of carbon dioxide hydrate phase equilibrium with and without ethanol. *Fluid Phase Equilibria,* v. 413, p. 176-183, 2016.

In: Pervaporation
Editor: Jean Garcia
ISBN: 978-1-53614-459-8

Chapter 2

The Fundamentals and Technology Surrounding Pervaporation Membranes

Tadashi Uragami*
Functional Separation Membrane Research Center, Osaka, Japan

Abstract

In this chapter, at first the definition, history and characteristics of pervaporation (PV) are described to understand the differences with other membrane separation technologies such as dialysis, ultrafiltration, microfiltration, nanofiltration and reverse osmosis. Furthermore, PV principle and PV fundamental analysis such as permeation equation are described. It is very important that we understand them in a judgment when we apply the membrane which we developed to the real application field.

In practical fields PV membrane materials: hydrophilic polymer materials, hydrophilic inorganic materials, hydrophobic polymer membrane materials, water permselective and organic permselective composite membranes, water-permselective, organic-permselective and organic-organic separation hybrid membranes, inorganic membranes, pervaporation technologies: dehydration such as water-alcohol dehydra-

* Corresponding Author Email: v701489@kansai-u.ac.jp.

tion and water-organic dehydration, organic separation from water: alcohol-selective concentration and organic selective removal, organic mixture separation such as separation of alcohol isomer, xylene mixtures, ether-alcohol mixture, hydrocarbon isomer, organic-organic azeotrope and desalination by PV, and technologies of wide field in PV are discussed in detail.

1. Definition and History of Pervaporation

This membrane permeation and separation method was named "Liquid Permeation" by Binning who did pioneer work in this field [1] and Choo named "Membrane Permeation" [2].

After the permeation and separation mechanism of this method was clarified, this method came to be called "Pervaporation (PV)," and was coined as a term for the combination of the *permeation* of the permeate through the membrane and its *evaporation* into the vapor phase based on the two basic steps of the process.

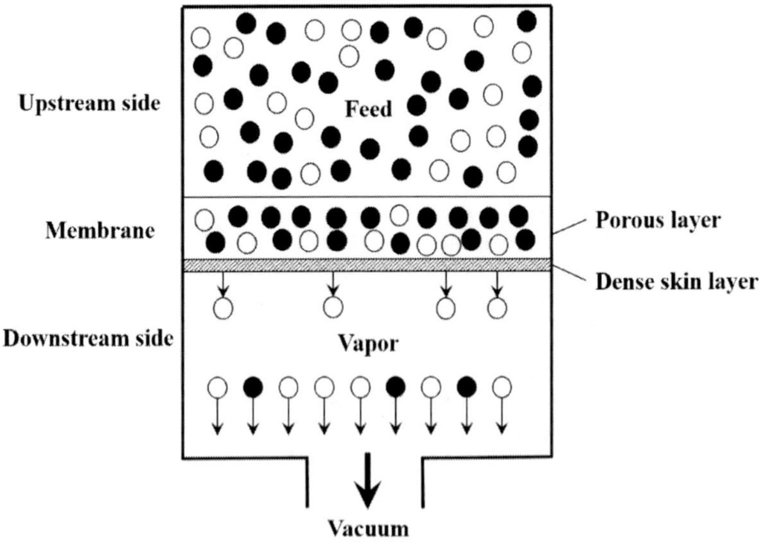

Figure 1. Principle of pervaporation.

PV membranes are generally composite structures which are consisted of a dense skin layer (active or permselective layer) and a porous layer (support layer) (Figure 1). Thosc membranes are effective for separation of organic liquid mixtures with high osmotic pressure and can be applied to the separation and concentration of azeotropic mixtures, close boiling point mixtures, and structural isomers, and can be used for the removal of certain components in equilibrium reactions.

The characteristics of PV are as follows:

1) A three-step process consisting of solution, diffusion, and evaporation, which significantly influences the selective transport across a non-porous membrane.
2) This process is effective for separation of organic liquid mixtures with high osmotic pressure because the driving force for permeation is the vapor pressure for each component rather than total system pressure.
3) The separation and concentration of mixtures that are difficult to separate by distillation are possible by PV. For example, it is useful for the separation of azeotropic mixtures, close- boiling-point mixtures, and structural isomers.
4) The removal of certain components in equilibrium reactions is enabled by PV.
5) Polymer membrane compaction, a frequent problem in high-pressure gas separations, is not encountered in PV because the feed pressure is typically low.

It's at the beginning of pervaporation separation that an aqueous solution of protein is kept in the bag of the semipermeable membrane and hung in the air, and the water is then vaporized and the solution is concentrated. Kahlenbergn studied the separation of hydrocarbon and alcohol by using a rubber membrane in 1906 although being qualitative [3]. The first quantitative study is the separation of an organic liquid azeotropic mixture udder high pressure using the microporous Vycor glass plate membrane [4]. Therefore, it was suggested that this method can be applied as one operation in the industry for the first time. Afterwards Heisler [5] in 1958 and Binnding et al. [1] in 1958 examined the possibility of the hybrid separation method with an organic homogeneous membrane. The study of Binning and others from the American Oil company yielded a satisfactory high separation rate and transmission speed with a very widespread thing for the time being. Furthermore, superior results from Michaels et al. [6] of MIT are presented, and they proposed the concept of a new liquid separation method. Choo, who is Michael's pupil, continuously studied and announced that pervaporation was an industrialization stage at mid-1960, but this process was not commercialized [2]. In the 1970s, Monsant picked up this process and assigned many patents covering a wide field of application in pervaporation [7, 8] but could not commercialize. Pervaporation research in the academic field was actively developed by Aptel and Neel group [9, 10], Cabasso [11, 12] and McCandless [13].

2. Fundamental Treatment in Pervaporation

2.1. Permeation Equation

At first, we consider permeation of the single component, not a mixed liquid. The permeation flux Q is expressed by Fick's first law as follows [14-18]:

$$Q = -D(C)\frac{dc}{dx} \tag{1}$$

where *c* is the concentration of solvent molecules in the polymer membrane, *x* is the membrane thickness. Fick's second law is as follows:

$$Q = -D(C)\frac{dc}{dx} \tag{2}$$

The expression used most commonly of this *D* (*C*) is often expressed by logarithmic relations, not a linear equation and a quadratic equation for *c*.

The diffusion coefficient depends on the concentration of material in the membrane, and is represented as follows:

$$D(C) = D_0 exp(aC) \tag{3}$$

where D_0 is the diffusion coefficient as the permeant concentration is zero, and a is the constant expressed as one standard of the plasticization action of the permeant. These two parameters depend on the permeation temperature, the crystallinity of the membrane, and the chemical and physical characteristics between the membrane material and the solvent molecule.

When equation 3 is substituted in equation 2 and is integrated using the boundary condition, where $C=C_1$ at $x=0$, $C=C_2$ at $x=l$, the equation 4 is obtained.

$$Q = \frac{D_0}{al}(\exp aC_1 - \exp aC_2) \tag{4}$$

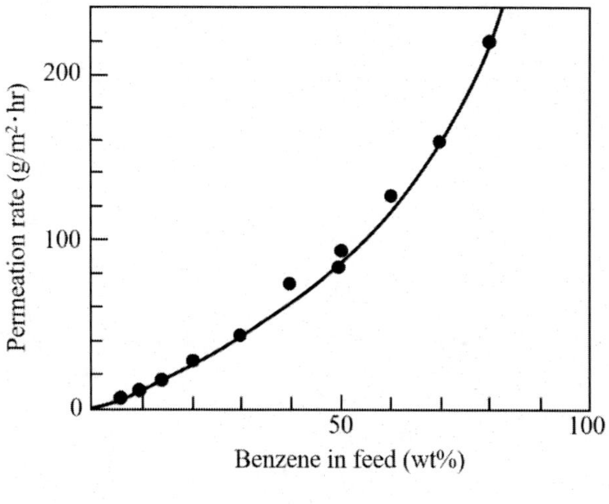

Figure 2. Effect of the feed composition on the total liquid permeation rate of benzene-aniline mixture (25°C) [19].

In other words, the permeation rate in the membrane is a logarithm to the solvent concentration in the membrane as seen in Figure 2. In addition, the concentration distribution in a membrane is as follows.

$$C = \frac{1}{a}\ln\left[\exp aC_1 - \frac{x}{l}(\exp aC_1 - \exp aC_2)\right] \tag{5}$$

When the concentration on the border of the membrane is at an equilibrium thermodynamically, Equation 6 is given.

$$C_1 = C^*(P^\circ), C_2 = C^*(P_2) \tag{6}$$

C^* is a functional relation, P^0 is the saturated vapor pressure of liquid at experimental temperature.

The permeability P is as follows:

$$\bar{P} = \frac{Ql}{\Delta P} = \frac{D_0}{a \cdot \Delta P}(\exp aC_1 - \exp aC_2) \tag{7}$$

where $\Delta P = P^0 - P_2$

In equation 6, when the Henry law is established, Equation 4, 5 and 7 are represented as a function of P^0 and P_2 like the following equations.

$$Q = \frac{D_0}{al}(\exp aSP^\circ - \exp aSP_2) \tag{8}$$

$$C = \frac{1}{a}\ln\left[\exp aSP^\circ - \frac{x}{l}(\exp aSP^\circ - \exp aSP_2)\right] \tag{9}$$

$$\bar{P} = \frac{D_0}{a\Delta P}(\exp aSP^\circ - \exp aSP_2) \tag{10}$$

These equations are proved experimentally.

We will think about the analytical method with the permeation ratio, Θ on analyzing the membrane transport separation behavior.

$$\Theta = Q/Q^\circ \tag{11}$$

Q is practical for all permeation rates of the mixture, Q^O is all permeation rate of mixture which is estimated from the permeation rate for each permeant. When the permeation ratio (θ_i) of each component is considered as an A, B two component system, it is defined as follows.

$$\theta_A = q_A/q_A{}^{\circ}, \theta_B = q_B/q_B{}^{\circ} \quad (12)$$

q_i and $q_i{}^0$ correspond to Q *and* Q^O of each component. As can be seen from these equations, if a system shows ideal behavior without the interaction between the components in the system, θ becomes unity. However, in general this is whether Q is bigger than unity, whether it is small. When θ is bigger than unity, this system increases the permeation rate, and if it is smaller than unity, the ideal permeation rate is depressed. Hamaya and Yamada [19] reported such results in aqueous dimethylamine solutions.

The permeation rate, Q through the membrane in a mixture of A and B is represented as follows:

$$Q = q_A + q_B \quad (13)$$

The membrane permeation in A and B mixture is the ideal state.

There is not an interaction between A and B, the permeation rate of each component is shown follows:

$$q_A = X_A \cdot q_A{}^{\circ}, q_B = X_B \cdot q_B{}^{\circ} \quad (14)$$

where X_A, X_B are the weight fractions of A and B components.

Overall permeation rate, Q^0 in an ideal system is

$$Q^{\circ} = X_A \cdot q_A{}^{\circ} + (1 - X_A)q_B{}^{\circ} \quad (15)$$

Also, the separation factor, $\alpha^0{}_{B/A}$ is as follows:

$$\alpha_{B/A} = \frac{Y_B/Y_A}{X_B/X_A} = \frac{q_B/q_A}{X_B/X_A} = \frac{X_B \cdot q_B{}^\circ}{X_A \cdot q_A{}^\circ} \cdot \frac{X_A}{X_B} = \frac{q_B{}^\circ}{q_A{}^\circ} \quad (16)$$

By integrating Equation 1

$$\mathrm{Q} \cdot l = -\int_{C_1}^{C_2} D(C)\,dc \quad (17)$$

where $Q = \bar{P}(P_1 - P_2)/l$ (18)

P_1 and P_2 are the vaper pressure in the membrane on the high and low concentration side respectively.

From Equations 17 and 18

$$Q = \bar{P}(P_1 - P_2)/l \quad (19)$$

By substituting Equation 19 for Equation 18, Equation (20) is obtained.

$$Ql \equiv R = \bar{P}(P_1 - P_2) = \int_{C_2}^{C_1} D(C)\,dc \quad (20)$$

where the concentration average diffusion coefficient, is defined as follows:

$$\bar{D} \equiv \left(\int_{C_2}^{C_1} D(C)\,dc\right)/(C_1 - C_2) \quad (21)$$

In Equation 21, if $\bar{D}$ does not depend to the concentration of permeant, it is equal with normal diffusion coefficient, *D*.

By substituting Equation (21) with Equation (20)

$$R = \bar{D}(C_1 - C_2) \quad (22)$$

$$\bar{P} = \bar{D}[(C_1 - C_2)/(P_1 - P_2)] \quad (23)$$

In many cases, because it is $P_1 \gg P_2$, Equations 21, 22 and 23 are omitted as follows:

$$\bar{D} = \int_0^{C_1} D(C)\,dc / C_1 \tag{24}$$

$$R \equiv \bar{D} C_1 \tag{25}$$

$$\bar{P} \equiv \bar{D}(C_1/P_1) \tag{26}$$

Here $C_1/P_1 = K_1$

Equation (27) is obtained.

$$\bar{P} = \bar{D} \cdot K_1 \tag{27}$$

K_1 is a pseudo solubility coefficient of the permeant on the high concentration side in the membrane and the concentration dependency. In conclusion, the permeability $\bar{P}$ is the product of the concentration average diffusion coefficient $\bar{D}$ and the solubility coefficient K_1. The concentration C_1 on the high concentration side in the membrane. Specific permeation rate, R is the product of the concentration average of diffusion coefficient $\bar{D}$ and the concentration C_1 on the high concentration side in the membrane.

3. Permeation and Separation Characteristics of Organic Liquid Mixtures

3.1. Water Selective Membranes for Water/Alcohol Mixtures

Water/alcohol selective membranes are effective for the following scenario: When an aqueous solution of dilute ethanol (about 10 wt %) produced by the bio-fermentation is concentrated by distillation, since an aqueous solution of 96.5 wt % ethanol is an azeotropic mixture, ethanol cannot be concentrated any more by distillation, and consequently ethanol is concentrated by azeotropic distillation with the addition of benzene. If membranes that can preferentially permeate only water at 3.5wt % in an azeotropic mixture of aqueous ethanol solution can be developed, significant energy savings would be achieved [20].

The permeation and separation mechanisms in PV through dense membranes consist of the dissolution of the permeants into the membrane, the diffusion of the permeants in the membrane, and the evaporation of the permeants from the membrane. Therefore, the separation of permeants in the membrane separation techniques depends on the differences in the solubility and diffusivity of the permeants in the feed mixture. When the structure of water/alcohol and water/organic liquid selective membranes are domically designed, hydrophilic materials can be recommended as membrane materials. Therefore, an increase in the solubility of water molecules into the membrane during the solution process can be expected.

To raise the affinity of membranes for water molecules, membranes with dissociation groups introduced into their structure are used for dehydration from organic solvents. Various membranes, which permeate water selectively from water/ethanol mixtures, were prepared by the plasma graft-polymerization of styrene (*g*-PS) onto porous poly (vinylidene fluoride) (PVF_2) films, by the sulfonation of the grafted membranes, and by the ionization of the sulfonated (*g*-PSS^- Na^+) membranes, respectively. The H_2O/EtOH selectivity increased with the ethanol concentration in the feed, and the grafted amount should be controlled for the optimum separation of water. The *g*-PSS^-Na^+ membrane with a grafted amount of PS of 0.14 mg/cm^2 was found to have a high permeation rate of 6.6 kg/ (m^2 h), and a separation factor of 21 for PV of aqueous 60 wt % ethanol solution at 50°C [21].

The PV separation of a concentrated ethanol/water mixture with 90 wt % of ethanol content through a sodium alginate (NaAgl) membrane was performed to investigate the permeation behavior of the membrane during PV [22]. From the swelling measurements of the membrane in an aqueous solution of 90 wt % ethanol, the solubility selectivity was about 1000 and the water content in the swollen membrane was 21 wt % at 40°C. Its excellent absorption properties could result in the outstanding PV performance for the aqueous solution; higher than a separation factor of 10,000 and permeation rate of 120-290 g/ (m^2 h).

Pervaporation separation of water–ethanol was carried out with poly-ion complex membranes based on κ-carrageenan [23]. The poly-ion complex membranes were prepared by the ion complex formation between k-carrageenan (anionic polymer) and poly{1,3-bis[4-alkylpyridi-nium]} propane bromides (cationic polymers) with different numbers of methylene units between two ionic sites within a repeating unit, respectively. The ion complex membranes were characterized with FT-IR, X-ray diffractometry. Dehydration of 90 wt % aqueous ethanol solution was carried out at different temperatures (30, 40, 50, and 60°C). The selectivity and permeability through them were very good over a wide temperature range; in the case of the poly-ion complex membrane consisting of k-carrageenan and poly {1,3-bis [4-ethylpyridinium] propane bromide}, the permselectivity was 45,000 and permeability was 150 g/m^2 h at 30°C. With increasing operating temperature, the permeability was increased highly but the selectivity decreased slightly.

Polyelectrolyte complex (PELC) membranes were prepared by simultaneous interfacial reaction of aqueous solutions of two oppositely charged poly-ions, i.e., from cellulose sulfate and various polycations as well as a cationic surfactant. Pervaporation investigations proved that such membranes prepared with polycations may be successfully used for dehydration of various organic solvents. Measurements of swelling and pervaporation properties of model membranes confirm, that the anionic polysaccharide Na-cellulose sulfate is the only component responsible for good separation capability in dehydration of organics with PELC membranes. Water molecules we assume to be preferentially transferred from one hydrogen bonding site to another across the polysaccharide chains. We can conclude that not the absorption into the upstream surface, but the diffusion selectivity along a swelling gradient across the stabilized cellulose sulfate mainly governs the separation behavior [24].

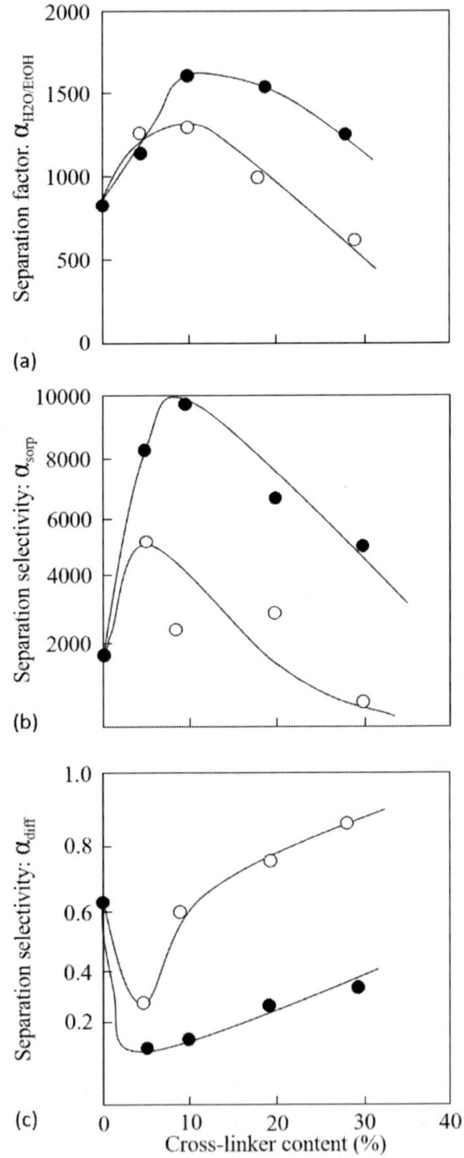

Figure 3 Effects of the cross-linker content on the separation factor, the sorption selectivity and the diffusion selectivity for an azeotrope of ethanol/water through CMCNa/GA cross-linked membranes (○) and CMCNa/TEOS hybrid membranes (●) during PV at 40°C [27].

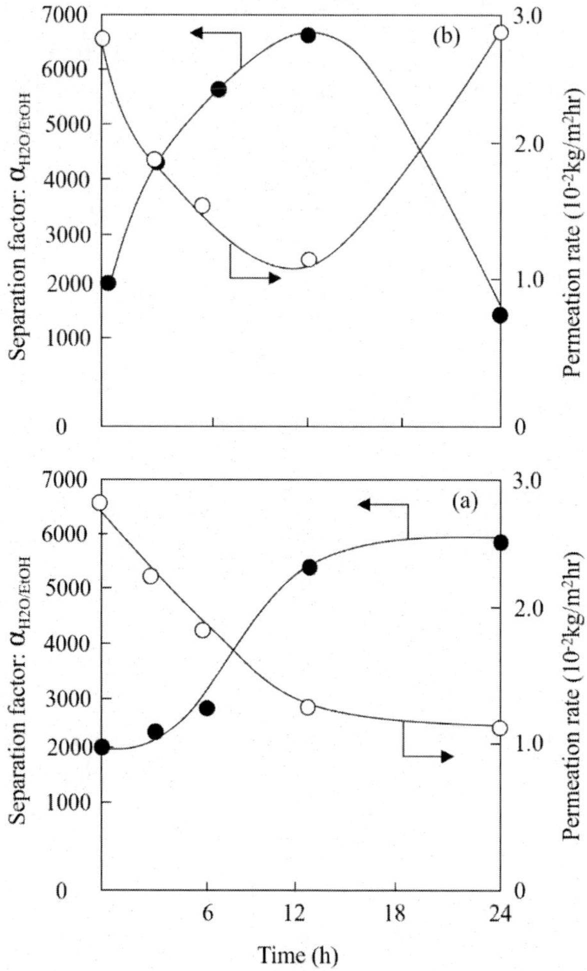

Figure 4. Effects of the immersion time in methanol solutions of $CaCl_2$ and $MgCl_2$ on the permeation rate (○) and separation factor (●) for an azeotropic mixture of ethanol/water through Alg-DNA/Ca^{2+} (a) and Alg-DNA/Mg^{2+} (b) cross-linked membranes during PV. The DNA content in each membrane is 40 wt % [28].

PIC membranes were prepared by the simultaneous interfacial reaction of aqueous solutions of two oppositely charged poly-ions, i.e., from cellulose sulfate and various polycations as well as a cationic surfactant. PV investigations proved that such membranes prepared with polycations may be successfully used for the dehydration of various organic solvents. Measurements of the swelling and PV properties of model membranes confirm that the anionic polysaccharide, Na-cellulose sulfate, is the only component responsible for good separation capability in the dehydration of organics with PIC membranes. PIC membranes consisting of chitosan (Chito) and poly (acrylic acid) (PAA) were prepared by blending two polymer solutions in different ratios. The thermal properties of PIC membranes constructed from chitosan and PAA by various blend ratios showed a shift in the transition temperature of the PICs.

PV performances were investigated with various organic mixtures; water/ethanol, water/1-propanol, and methanol/methyl *t*-butyl ether mixtures. An increase in the PAA content of PIC membranes affected the swelling behavior and PV performance of a water/ethanol mixture. The permeation rate decreased, and the water concentration in the permeate was close to 100% upon increasing the feed alcohol concentration [25]. Asymmetric PIC membranes composed of chitosan membrane and PAA were constructed from an aqueous solution. The absorption quantity of PAA decreased, whereas the water selectivity of the PIC membrane increased with an increase in the molecular weight of PAA. In PV experiments, the water selectivity of the membrane was so high that no ethanol was detected by gas chromatography [26].

To control the swelling of sodium carboxymethylcellulose (CMCNa) membranes, mixtures of CMCNa and glutaraldehyde (GA) and mixtures of CMCNa as an organic component and tetraethoxysilane (TEOS) as an inorganic component were prepared, and CMCNa/GA cross-linked membranes and CMCNa/TEOS hybrid membranes were formed. In the separation of an ethanol/water azeotrope by pervaporation (PV), the effects of the GA or TEOS content on the water/ethanol selectivity and permeability of these CMCNa/GA cross-linked and CMCNa/TEOS hybrid membranes were investigated. Cross-linked and hybrid membranes containing lip to 10 wt % GA or 10 wt % TEOS exhibited higher water/ethanol selectivity than CMCNa membrane without any CMCNa/TEOS hybrid membranes and their pe cross-linker. This resulted from both increased density and depressed swelling of the membranes by the formation of a cross-linked structure. The relationship between the structure of the CMCNa/GA cross-linked membranes and CMCNa/TEOS hybrid membranes and their permeation and separation characteristics for an ethanol/water azeotrope during PV is discussed in detail [27].

In Figure 3, the separation factor, the absorption selectivity and the diffusion selectivity for an azeotrope of ethanol/water through CMCNa/GA cross-linked or CMCNa/TEOS hybrid membranes are shown as a function of the GA or TEOS content. The absorption selectivity of all membranes with various GA or TEOS contents were greater than their diffusion selectivity. This observation suggests that the dehydration process from an aqueous solution of 96.5 wt % ethanol using CMCNa/GA cross-linked and CMCNa/TEOS hybrid membranes is mainly governed by the absorption process. The fact that water was preferentially absorbed into these membranes rather than ethanol also supports this conclusion. Furthermore, the absorption selectivity of the CMCNa/TEOS hybrid membranes were higher than those of the CMCNa/GA cross-linked membranes. These results are supported by the composition absorbed into the membrane. On the oilier hand, the diffusion selectivity of the former membranes were smaller than those of the latter membranes. This is due to the fact that with increasing GA or TEOS content, the degree of swelling of the membrane decreased as shown in Figure 3. The increase in the water permselectivity with increasing GA or TEOS content was mainly due to an increase in the solubility of water in the CMCNa/GA cross-linked and CMCNa/TEOS hybrid membranes.

To obtain high dehydration membranes for ethanol/water azeotrope. dried blend membranes prepared from mixtures of sodium alginate (Alg-Na) and sodium deoxyribonucleate (DNA-Na) were cross-linked by immersing in a methanol solution of $CaCl_2$ or $MaCl_2$. In the dehydration of an ethanol/water azeotropic mixture by pervaporation, the effects of immersion time in methanol solution of $CaCl_2$ or $MaCl_2$ on the permeation rate and water/ethanol selectivity through Alg-DNA/Ca^{2+} and Alg-DNA/Mg^{2+} cross-linked membranes were investigated. Alg-DNA/Mg^{2+} cross-linked membrane immersed for 12 hours in methanol solution of $MaCl_2$ exhibited the highest water/ethanol selectivity. This results from depressed swelling of the membranes by formation of a cross-linked structure. However, excess immersion in solution containing cross-linker led to an increase in the hydrophobicity of cross-linked membrane. Therefore, the water/ethanol selectivity of Alg-DNA/Mg^{2+} cross-linked membranes with an excess immersion in cross- linking solution was lowered. The relationship between the structure of Alg-DNA/Ca^{2+} and Alg-DNA/Mg^{2+} cross-linked membranes and their permeation and separation characteristics during pervaporation of an ethanol/water azeotropic mixture is discussed in detail [28].

In Figure 4, the permeation rate and separation factor for an ethanol/water azeotrope through the Alg-DNA/Ca^{2+} and Alg- DNA/Mg^{2+} cross-linked membranes during PV are as a function of the immersion time in the CH_3OH solutions of $CaCl_2$ and $MgCl_2$. In Alg-DNA/Ca^{2+} cross-linked membranes in Figure 4 (a), with increasing the immersion time the separation factor increased and the permeation rate decreased. All separation factors were much higher than those in the Alg-DNA blend membrane without the cross-linking. These results suggest that the Alg-DNA/Ca^{2+} cross-linked membranes showed much higher H_2O/EtOH selectivity.

The poly-ion complex membranes were prepared by ion complex formation (PIC) between κ-Carrageenan and poly {1,3-bis [4-alkypyridinium] propane bromides}, with different numbers of methylene units between the two ionic sites within a repeating unit, respectively. The dehydration of a 90 wt % aqueous ethanol solution was carried out at different temperatures during. The selectivity and permeability were very good over a wide temperature range; in the case of the PIC membrane consisting of κ-carrageenan and poly {1,3-bis [4-ethylpyridinium] propane bromide}, the H_2O/EtOH selectivity was 45,000 and the permeability was 150 g/(m^2h) at 30°C. With increasing operating temperature, the permeability increased greatly but the selectivity decreased slightly [29]. The poly-ion complex PIC membranes were prepared by the simultaneous interfacial reaction of aqueous solutions of two oppositely charged poly-ions, i.e., from cellulose sulfate and various polycations as well as a cationic surfactant. PV investigations proved that such membranes prepared with polycations may be successfully used for the dehydration of various organic solvents. Measurements of the swelling and PV properties of model membranes confirm that the anionic polysaccharide, Na-cellulose sulfate, is the only component responsible for good separation capability in the dehydration of organics with PIC membranes [30]. PIC membranes consisting of chitosan (Chito) and poly (acrylic acid) (PAA) were prepared by blending two polymer solutions in different ratios.

The thermal properties of PIC membranes constructed from chitosan and PAA by various blend ratios showed a shift in the transition temperature of the PICs. PV performances were investigated with various organic mixtures; water/ethanol, water/1-propanol, and methanol/methyl *t*-butyl ether mixtures. An increase in the PAA content of PIC membranes affected the swelling behavior and PV performance of a water/ethanol mixture. The permeation rate decreased, and the water concentration in the permeate was close to 100% upon increasing the feed alcohol concentration [31]. Asymmetric PIC membranes composed of chitosan membrane and PAA were constructed from an aqueous solution. The absorption quantity of PAA decreased, whereas the water selectivity of the PIC membrane increased with an increase in the molecular weight of PAA. In PV experiments, the water selectivity of the membrane was so high that no ethanol was detected by gas chromatography [32].

The dehydration of aqueous alcohol solutions through asymmetric nylon 4 (N4) membranes was investigated using the PV processes. A separation factor of 4.72 and a permeation rate of 0.78 kg/(m^2h) for the asymmetric membrane were obtained. When compared to conventional homogeneous N4 membranes, the asymmetric membrane can effectively supersede the PV performance of the N4 membrane for the separation of water/alcohol mixtures [33]. A 4-vinylpyridine-grafted-polycarbonate (PC-*g*-4VP) membrane was prepared for PV of water selectivity. Water was permeated through the PC-*g*-4VP membrane preferentially over all feed compositions. The permeation rate and separation factor increased with increasing 4-vinylpyridine content in the membrane. The total permeation rate of a PC-*g*-4VP membrane with grafting of 26.7% was 153 g/ (m^2 h), and the separation factor was above 6300 for an aqueous solution of 90 wt % ethanol [34]. A hydrophilic PV membrane was prepared via the homograft polymerization of *N, N'*-(dimethyl amino) ethyl methacrylate (DMAEM) onto a nylon N4 backbone, thus generating the DMAEM-*g*-N4 membrane [35]. The water selectivity was improved by the ammonium quaternization of the pendant *N, N'*-dimethyl amino group on the DMAEM-*g*-N4 membrane using dimethyl sulfate, resulting in a DMAEMQ-*g*-N4 membrane. The separation factor and permeation rate for both chemically modified N4 membranes were higher than those of the unmodified N4 membrane.

Optimum PV was obtained by a DMAEMQ-*g*-N4 membrane with a degree of grafting of 12.7% for a 90 wt % ethanol feed concentration, giving a separation factor of 36 and a permeation rate of 564 g/ (m^2 h). Temperature-sensitive membranes were synthesized by grafting poly (*N*-isopropyl-acrylamide) (PNIPAAm) onto a PVA backbone using hydrogen peroxide-ferrous ion as an initiator. Due to the grafting of PNIPAAm, the hydrophilic/ hydrophobic balance and the polarity of the pendant groups within the membranes were modified. Significant temperature sensitivity of the grafted membranes was observed close to the lower critical solution temperature (LCST) of linear PNIPAAm membranes during the PV of an ethanol/water mixture. Both the PV and absorption selectivity for water showed a maximum value near 30-32°C for ethanol contents of 75 and 80%. The temperature sensitivity of the grafted membranes also depended on the ethanol concentration. The maximum PV and solubility selectivity disappeared when the ethanol content was lower than 75%, because the greater degree of swelling reduced the size screening effect of the membrane [36].

Interpenetrating polymer network (IPN) PV membranes were prepared by the free-radical polymerization of acrylamide (AAm) or acrylic acid (AA) in the presence (or absence) of the cross-linking agents allyl dextran or *N, N'*-methylenebisacrylamide within cellophane films swollen in the reaction mixture. IPN membranes were selective over a wide range of ethanol concentrations in the feed. The separation factor and the permeation rate improved significantly with increasing PAAm-PAA in the IPN membranes, especially for the cellulose-PAA (K^+ form) membranes; for an 86% EtOH feed at 50°C. The separation characteristics of these membranes were in good correlation with their swelling behavior [37].

Three different types of blend membranes based on Chito and PAA were prepared, and their performance for the PV separation of a water/ethanol mixture was investigated. All membranes were highly H_2O/EtOH selective. The temperature dependence of the membrane for H_2O/EtOH selectivity for feed solutions of higher water content (> 30 wt %) was unusual in that both the permeability and the separation factor increased with increasing temperature. A comparison of the PV performance between composite and homogeneous membranes was made, and typical PV results at 30°C for a 95 wt % ethanol aqueous solution were as follows:

For the homogeneous membrane, the permeation rate was 33 g/(m^2·h), and the separation factor was 2216; and for the composite membrane, the permeation rate was 132 g/(m^2·h), and the separation factor was 1,008 [38]. Blend membranes were prepared by coagulating a mixture of *O*-carboxy-methylated chitosan (CM-Chito) and alginate in an aqueous solution with 5 wt % $CaCl_2$, and then by treating it with a 1 wt % HCl aqueous solution. Polymer interpenetration including a Ca^{2+} cross-linked bridge occurred in the blend membrane and resulted in a high separation factor for the PV separation of alcohol/water and low permeation rate. The thermostability of the blend membranes was significantly superior to that of alginic acid and cellulose/alginate blend membranes, owing to a strong electrostatic interaction caused by the amino groups of CM-Chito with the carboxylic acid groups of alginic acid [39].

Three different types of blend membranes based on chitosan and polyacrylic acid were prepared from homogeneous polymer solution and their performance on the pervaporation separation of water-ethanol mixtures was investigated. It was found that all membranes are highly water-selective. The temperature dependence of membrane permselectivity for the feed solutions of higher water content (>30 wt %) was unusual in that both permeability and separation factor increased with increase in temperature. This phenomenon might be explained from the aspect of activation energy and suggested that the absorption contribution to activation energy of permeation should not always be ignored when strong interaction occurs in the pervaporation membrane system.

A comparison of pervaporation performance between composite and homogeneous membranes was also studied. Typical pervaporation results at 30°C for a 95 wt % ethanol aqueous solution were: for the homogeneous membrane, permeation flux = 33 g/ (m^2 h), separation factor = 2216; and for the composite membrane, permeation flux = 132 g/m^2 h, separation factor = 1008. A transport model consisting of dense layer and porous substrate in series was developed to describe the effect of porous substrate on pervaporation performance [40].

The blend membranes were satisfactorily prepared by coagulating a mixture of O-carboxymethylated chitosan (CM-chitosan) and alginate in aqueous solution with 5 wt % $CaCl_2$, and then by treating with 1 wt % HCl aqueous solution. Their structure and miscibility were characterized by scanning electron micrograph, X-ray diffraction, infrared spectra, differential thermal analysis, and atomic absorption spectrophotometer. The results indicated that the blends were miscible, when the weight ratio of CM-chitosan to alginate was in the range from 1:1 to 1:5. The polymers interpenetration including a Ca^{2+} crosslinked bridge occurred in the blend membrane, and leads to high separation factor for pervaporation separation of alcohol/water and low permeation. The tensile strength in the wet state (σ_b = 192 kg cm^{-2} for CM-chitosan/alginate 1:1) and thermostability of the blend membranes were significantly superior to that of alginic acid membrane, and cellulose/alginate blend membranes, owing to a strong electrostatic interaction caused by —NH_2 groups of CM-chitosan with —COOH groups of alginic acid [41].

The structure of chitosan membranes chemically modified with aldehydes, such as glutaraldehyde and *n*-butyl aldehyde, was analyzed. Characteristics of permeation and separation of aqueous ethanol solutions through the membranes were investigated by evaporation. Chitosan membranes reacted with glutaraldehyde were cross-linked by Schiff base bonds and had no pendant structure. Density and crystallinity of the cross-linked chitosan membranes decreased with increasing glutaraldehyde content in the membranes. The permeation rate, separation factor for water permselectivity, and degree of swelling of the cross-linked chitosan membrane for aqueous ethanol solutions increased with increasing glutaraldehyde content in the chitosan membrane. The structure of the chemically modified chitosan membrane and the permeation and separation characteristics for aqueous ethanol solutions are discussed from the viewpoint of hydrogen bonds of the cross-linked membrane [42].

To obtain water permselective membranes, the introduction of dissociation groups and hydrophilic groups into polymer membrane materials, chemical modification, graft copolymerization, and polymer blends have all been attempted. However, adequate membrane performance cannot be obtained by only improving hydrophilicity of the membrane materials. Most importantly, membrane performance is significantly affected by the swelling of the polymer membrane due to the feed mixture. Cross-linking is one of methods that can maintain the hydrophilicity of the membrane while attenuating the swelling of the membrane.

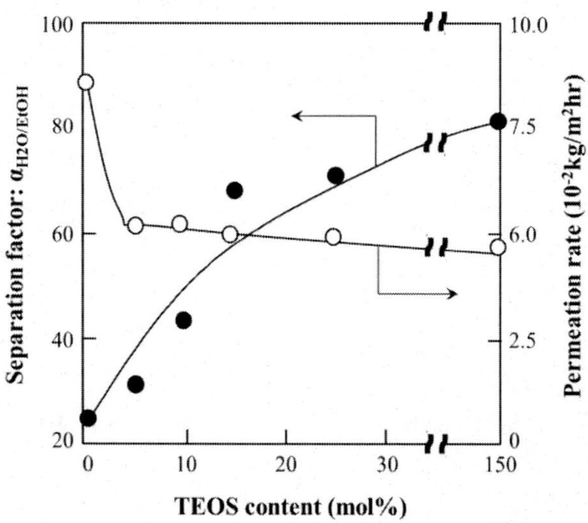

Figure 5. Effects of the TEOS content in PVA on the permeation rate (●) and the separation factor for the water permselectivity (○) of an aqueous solution of 86 wt % ethanol through PVA-TEOS hybrid membranes by pervaporation at 40°C.

Organic-inorganic hybrid membranes composed of PVA and TEOS were prepared as the dehydration membrane of an aqueous ethanol solution in pervaporation. The effect of the TEOS content in the PVA-TEOS hybrid membranes on the permeation rate and separation factor for the water permselectivity of an aqueous solution of 85 wt % ethanol during pervaporation is shown in Figure 5. As can be seen in this figure, the separation factor for water permselectivity increased with increasing TEOS content, but the permeation rate became constant after a decrease at low TEOS content [43].

To inhibit the swelling of PVA membranes in aqueous solutions, which leads to lowered water permselectivity during separation, organic-inorganic hybrid membranes composed of PVA and tetraethoxysilane (TEOS) were prepared. When an aqueous ethanol solution was permeated through the PVA/TEOS hybrid membranes during PV, the H_2O/EtOH selectivity increased, but the permeation rate decreased, with increasing TEOS content. This decreased permeation rate caused a decreased degree of swelling of the membrane. This decrease in the degree of swelling and increase in the membrane density were due to the formation of hydrogen bonds between the silanol groups resulting from the hydrolysis of TEOS and the hydroxyl group of PVA. When the PVA and PVA/TEOS hybrid membranes were annealed, the H_2O/EtOH selectivity of these membranes increased with increasing annealing temperature and time. The fact that annealing at higher temperatures promoted the dehydration-condensation reaction between PVA and TEOS in the PVA/TEOS membranes was related to the enhanced H_2O/EtOH selectivity of the PVA/TEOS membranes [43]. Hydrophilic organic-inorganic hybrid membranes were prepared from hydrophilic *q*-Chito and TEOS by a sol-gel process, in order to minimize the swelling of the *q*-Chito membranes. When an azeotrope of ethanol/water was permeated through their *q*-Chito/TEOS hybrid membranes during PV, the *q*-Chito/TEOS hybrid membranes showed high H_2O/EtOH selectivity. However, the H_2O/EtOH selectivity of the membranes decreased slightly with increasing TEOS content over 45 mol %. Furthermore, the H_2O/EtOH selectivity of these membranes is discussed from the viewpoint of their chemical and physical membrane structure [44].

Swelling of poly (vinyl alcohol-co-acrylic acid) (P(VA-*co*-AA)) membranes in aqueous alcohol solutions operated under PV conditions leads to low H_2O/EtOH selectivity. To reduce swelling, organic-inorganic hybrid membranes composed of P(VA-*co*-AA) and tetraethoxysilane (TEOS) were prepared. However, when an aqueous ethanol solution was permeated through the P(VA-*co*-AA)/TEOS hybrid membranes by PV, the permeation rate increased and the H_2O/EtOH selectivity decreased with increasing TEOS content. The increase in the permeation rate and the decrease in the H_2O/EtOH selectivity were caused by an increase in the degree of swelling of the membrane and a decrease in the membrane density with increasing TEOS content. These effects resulted from insufficient formation of hydrogen bonds between the silanol groups by hydrolysis of TEOS and the hydroxyl and carboxyl groups of P(VA-*co*-AA). When the P(VA-*co*-AA)/TEOS hybrid membranes were annealed, the water/ethanol separation factor for H_2O/EtOH selectivity increased with increasing annealing time and TEOS content. Longer annealing time promoted the dehydration-condensation reaction between P(VA-*co*-AA) and TEOS in P(VA-*co*-AA)/TEOS hybrid membranes, leading to enhanced H_2O/EtOH selectivity of the hybrid membranes [45].

To control the swelling of PVA membranes, mixtures of PVA and an inorganic polysilane were prepared using sol-gel reactions to yield new PVA/polysilane hybrid membranes. In the separation of an ethanol/water azeotropic mixture during PV, the effect of the polysilane content on the H_2O/EtOH selectivity of PVA/polysilane hybrid membranes were investigated. The H_2O/EtOH selectivity of PVA/polysilane hybrid membranes were higher than that of PVA membranes, but the H_2O/EtOH selectivity of hybrid membranes decreased with increasing polysilane content. To increase the H_2O/EtOH selectivity, PVA/polysilane hybrid membranes were annealed. The H_2O/EtOH selectivity of annealed PVA/polysilane hybrid membranes were greater than annealed hybrid membranes, and significantly governed by the content, which could be attributed to both solubility and diffusion selectivity. The relationship between the structure of un-annealed and annealed PVA/Polysilane hybrid membranes along with permeation and separation characteristics of an ethanol/water azeotropic mixture during PV are discussed in detail [46]. PV and EV separations of water/alcohol mixtures through a modified N4 membrane- coated with plasma deposited, hydrolyzed vinyl acetate (VAc) - (PVA-*p*-N4), were investigated. The separation factor and the permeation rate of this PVA-*p*-N4 membrane are both higher than those of the unmodified N4 membrane for PV of aqueous ethanol solutions. The PVA-*p*- N4 membrane showed a separation factor of 13.5 and a permeation rate of 420 g/ (m^2 h) can be obtained. Compared with PV, EV effectively increased the separation factor for the water/alcohol mixtures, but the permeation rate decreased the permeation rate [47].

To improve the hydrophilicity of a N 4 membrane for PV and EV processes, and to overcome the hydrolysis of PVA, a PVA-*g*-N4 membrane by γ-ray irradiation grafting of vinyl acetate (VAc) onto N4 membrane was prepared, and then followed by hydrolysis treatment. A separation factor of 13.8 and a permeation rate of 0.352 kg/ (m^2 h) can be obtained for a PVA-*g*-N4 membrane with a degree of grafting of 21.2% for a 90 wt % ethanol feed concentration. Compared to the PV process, the EV process had a significantly increased separation factor with a decreased permeation rate for the same PVA-*g*-N4 membrane [48].

Twelve kinds of polyimide membranes were prepared using three dianhydrides (including 2,2-bis[4-(3,4-dicarboxyphenoxy)phenyl]propane dianhydride (BPADA), 3,3'4,4'-benzophenonetetracarboxylic dianhydride (BTDA), and 3,3'4,4'-biphenyltetracarboxylic dianhydride (ODPA)) and four diamines (including benzidine (BZD), bis(4-aminophenyl)phenyl phosphate (BAPP), 4,4'-diaminodiphenylmethane (MDA), and 4,4'-diaminodiphenyl ether (ODA)) via a two-step method. The permeation rate of ethanol/water mixtures through the polyimide membranes with the same dianhydrides increases following the order of BZD < ODA < MDA < BAPP. The permeation rate increases with increase in temperature [49].

Zeolite-embedded hybrid membranes are manufactured by using a casting machine with polyester nonwoven fabric as the supporting layer, polyacrylonitrile (PAN) as the porous backing layer and PVA as the active separating layer. Experimental results show the H_2O/EtOH selectivity has been greatly improved after adding zeolite 4A and that reasonably high separation factor can be achieved for feed ethanol concentration of above 80 wt %, probably due to the superior molecular sieving effect of added zeolite 4A on the water/ethanol system. By incorporation of zeolite, apparent Arrhenius activation energy significantly decreased for water but obviously increased for ethanol. Water molecules require much less energy whereas ethanol molecules need much more energy to transport through the membrane because of the hydrophilic characteristics of zeolite 4A [50].

The separation of aqueous alcohol mixtures was carried out by use of a series of novel aromatic polyamide membranes. The aromatic polyamides were prepared by the direct polycondensation of 2,2'-dimethyl-4,4'-bis(aminophenoxyl)biphenyl (DBAPB) with various aromatic diacids, such as terephthalic acid (TPAc), 5-tert-butylisophthalic acid (TBPAc), and 4,4'-hexafluoroisopropylidenedibenzoic acid (FDAc). The solubility of ethanol in the aromatic polyamide membranes is higher than that of water, but the diffusivity of water through the membrane is higher than that of ethanol. The effect of diffusion selectivity on the membrane separation performances plays an important role in the EV process. Compared with PV, EV effectively increased the H_2O/EtOH selectivity. Moreover, the effect of aromatic diacids on the polymer chain packing density on the PV, and EV performance were investigated. The permeation rate could be increased by introduction of a bulky group into the polymer backbone [51].

A series of soluble polyimides derived from 3,3',4,4'-benzhydrol tetracarboxylic dianhydride (BHTDA) with various diamines such as 1,4-bis(4-aminophenoxy)-2-tert-butylbenzene (BATB), 1,4-bis(4-aminophenoxy) 2,5-di-tert-butylbenzene (BADTB), and 2,2'-dimethyl-4,4'- bis(4-aminophenoxy)biphenyl (DBAPB) were investigated for PV separation of ethanol/water mixtures. Diamine structure effect on the PV of 90 wt % aqueous ethanol solution through the BHTDA-based polyimide membranes was studied. The H_2O/EtOH selectivity was ranked in the order of BHTDA-DBAPB>BHTDA-BATB>BHTDA-BADTB. The increase in molecular volume for the substituted group in the polymer backbone increased the permeation rate. As the feed ethanol concentration increased, the permeation rate increased, while the water concentration in the permeate decreased for all polyimide membranes. The optimum PV performance was obtained by the BHTDA-DBAPB membrane with an aqueous ethanol solution of 90 wt %, giving a separation factor of 141, permeation rate of 255 g/(m^2 h) and 36,000 of PSI [52].

Novel organic-inorganic hybrid membranes were prepared through sol-gel reaction of PVA with γ-aminopropyltriethoxysilane (APTEOS) for PV separation of ethanol/water mixtures. The amorphous region of the hybrid membranes increased with increasing APTEOS content, and both the free volume and the hydrophilicity of the hybrid membranes increased when APTEOS content was less than 5 wt %. The swelling degree of the hybrid membranes has been restrained in an aqueous solution owing to the formation of hydrogen and covalent bonds in the membrane matrix. Permeation rate increased remarkably with APTEOS content increasing, and H_2O/EtOH selectivity increased at the same time, the trade-off between the permeation rate and H_2O/EtOH selectivity of the hybrid membranes was broken. The absorption selectivity increased with increasing temperature but decreased with increasing water content. The hybrid membrane containing 5 wt % APTEOS has the highest separation factor of 536.7 and permeation rate of 0.0355 kg/ (m^2 h) in PV separation of 5 wt % water in the feed [53].

Sericin/PVA blend membranes were prepared by blending sericin and PVA, followed by chemical cross-linking with dimethylurea. The blend membranes were preferentially permeable to water. In the temperature range of 50-70°C, a permeate water concentration of 93.1-94.1 wt % was achieved at 8.5 wt % water in the feed. As a comparison, membranes were also fabricated from pure sericin and PVA alone, and tested for PV separation under the same conditions. The water selectivity of membrane was primarily derived from absorption selectivity, and there was a strong coupling effect for the permeation and absorption of the permeant in the membranes [54].

Energy efficient dehydration of low water content ethanol is a challenge for the sustainable production of fuel-grade ethanol. Blends of cross-linked mixture of poly (allylamine hydrochloride) and hydrolyzed PVA of 99 and 88% were prepared as pervaporative dehydration membrane. These polymeric membranes had higher HO/EtOH selectivity for low feed water concentrations (< 5 wt %) than 99% hydrolyzed PVA [55].

Novel semi-IPN membranes of sodium alginate (NaAlg) and thermo-responsive PNIPAAm were prepared by free radical polymerization using potassium persulfate as an initiator [56]. Membranes were cross-linked with GA and used in PV separation of water/ethanol mixtures at varying feed compositions and temperatures. PV separation characteristics of the membranes showed a dependence on thermosensitive nature of NIPAAm.

Increasing the NIPAAm content of the semi-IPN network resulted in an increased selectivity with a decreased permeation rate. The membrane containing 30 wt % NIPAAm showed the highest H_2O/EtOH selectivity of 18,881 with a permeation rate of 0.137 kg/(m^2 h) at 40°C, i.e., the above LCST of NIPAAm, but below its LCST, i.e., at 25°C, it exhibited considerably a lower H_2O/EtOH selectivity of 92 with a permeation rate of 0. 185 kg/($m^{2/}$h) for 15 wt % water in the feed. The H_2O/EtOH selectivity decreased with decreasing temperature at 15 wt % water below the LCST region, i.e., at 30°C. The pore size of the membrane became larger, the permeation rate increased up to 0.225 kg/ (m^2 h), but the H_2O/EtOH selectivity decreased.

Sulfonated poly (phenylene oxide) (SPPO) membranes were prepared for dehydrating water/ethanol mixtures. The effects of hydration of sulfonated membranes on PV performance were discussed by comparing the characteristics of pure PPO and SPPO membranes. The differences in microstructural and hydrophilic properties of membranes were characterized by means of atomic force microscopy and a swelling test. The degree of sulfonation of PPO significantly affected the hydrophilicity of sulfonated membranes and played an important role for the dehydration performance of PV membranes. The sulfonated membranes showed excellent water permeation rate of about 300 g/ (m^2 h) with good selectivity up to 700. SPPO membranes exhibited better PV performance than pure PPO membranes [57].

The asymmetric aluminum ion exchange polysulfone membranes were prepared for the dehydration of ethanol/water mixture. The separation performance of those membranes was increased with increasing the degree of aluminum ion exchange in polysulfone membranes. Both the permeation rate and separation factor of those membranes increased with increasing the degree of ion exchange. The increase in separation performance of aluminum ion exchange membranes was mainly attributed to ion cross-linking in polymer network and the degree of hydration of exchanged ion in membranes. On the other hand, the operating temperature in the PV process showed a significant influence on the dehydration of water molecules in the permeate. An increase in temperature increased the permeation rate of permeate but slightly decreased H_2O/EtOH selectivity [58].

Through the complexation of two anionic polysaccharide blends composed of NaAlg and carrageenan with divalent calcium ions, PIC membranes were prepared. The effects of annealing on the structure of the PIC membranes and on their performance at removing water from a methanol mixture were investigated. The annealed membranes exhibited a change in their crystallinity. The change was due to a rearrangement of polymer chains, which was induced by a deformation of the chelate structure and intramolecular or intermolecular interactions between the polysaccharides. Because of the effects of annealing on the resulting membranes, water components almost penetrated the membranes to the permeate side in the vapor permeation process. Moreover, the membrane performances were gradually enhanced as the operating temperature increased [59].

Cross-linked, dense PVA membranes with different degrees of hydrolysis were prepared and used in the absorption and PV of 2-propanol (IPA)/water mixtures. The partial permeation rate for water permeation was increased with increasing water content in the liquid mixture, but the partial permeation of 2-propanol due to the coupling effect of absorption and permeation reached a maximum value. The degree of PVA hydrolysis and the feed temperature influenced the permeation rate and H_2O/IPA selectivity due to the degree of PVA hydrolysis, and the selectivity of PVA for water was inversely proportional to the degree of PVA hydrolysis [60].

Homogeneous cellulose triacetate membranes were prepared by the solution-casting method. The surface of this membrane was modified with gaseous plasma of a 10-W discharge power in the presence of ammonia gas at 0.15 Torr pressure. The percentage of weight loss of the CTA membranes was found to be 0.7 for 20 min of treatment time in the ammonia plasma. The contact angle measurement indicated that hydrophilicity of the surface increases. ATR-FTIR spectral analysis showed that the hydrophilicity is mainly derived from the amino groups on the modified surface. SEM studies indicate that no considerable change of surface morphology occurred up to 5 min of treatment time, but a considerable change of surface morphology resulted for treatment of 10 and 20 min. The modified membranes were used for pervaporation studies for separation of an isopropanol–aqueous mixture. These membranes showed excellent selectivity for water. The water flux increases with an increase in treatment time for all concentrations of isopropanol in the feed. The isopropanol flux decreases for initial treatment time (2 and 5 min), but showed an increasing trend for a higher treatment time (10 and 20 min) [61].

Surface modification of cellulose acetate (CTA) membranes was achieved by subjecting them to magnetron-enhanced plasma. Plasma polymerization and deposition of thiophene were achieved on these membranes in a nitrogen atmosphere. Such modified membranes were characterized by determining the rate of deposition, angle of contact with water, and IR spectroscopy. The morphology of deposition was studied by SEM, and it was found that the globular particles grow with time to give a honeycomb or platelet structure. The water vapor transmission rates were measured and were found to decrease with an increase in the time of deposition. The pervaporation process of such modified CTA films for the separation of an isopropanol–water mixture was investigated. The water selectivity of the modified membranes was found to be increased [62].

The surface of a cellulose triacetate membrane was modified with gaseous plasma in the presence of ammonia gas. The contact angle measurement indicated that the hydrophilicity of the surface increased. These membranes showed excellent H_2O/IPA selectivity for water in the separation of an aqueous solution of 2-propanol. The permeation rate of water increased with an increase in the treatment time for all concentrations of IPA in the feed. Surface modified cellulose acetate membranes treated with magnetron-enhanced plasma also showed high H_2O/IPA selectivity for the separation of a 2-propanol/water mixture during PV.

Blend membranes of Na-Alg and PAAm grafted guar gum (PAAm-*g*-GG) in ratios of 3:1 and 1:1 were prepared for the PV separation of a water/2-propanol mixture. Membranes prepared from pure Na-Alg (M-1) and 1:1 blends of Na-Alg and PAAm-g-GG (M-3) showed the highest separation selectivity for 10 mass % water in the feed mixture, whereas a membrane prepared with a 3:1 blend ratio of Na-Alg and PAAm-g-GG showed the highest permeation selectivity for 20 mass % water in the feed.

The H_2O/IPA selectivity decreased with increasing amounts of grafted copolymer in the blend mixture. The permeation rate increased with increasing amounts of water in the mixture, but the permeation rate did not change markedly with the PAAm-*g*-GG content in the blend membrane at the lower mass % water [63].

Cross-linked organic-inorganic hybrid Chito membranes were obtained from blending Chito and γ-(glycidyloxypropyl) trimethoxy-silane (GPTMS) in acetic acid aqueous solution. The hydrophilicity of the modified membranes was not significantly decreased, to result in good H_2O/IPA selectivity and high permeation rate in pervaporative dehydration on a 70 wt % 2-propanol/water mixture. The chitosan membrane containing 5 wt % GPTMS had a permeation rate of 1,730 g/ (m^2 h) and separation factor of 694 for H_2O/IPA selectivity [64].

The effectiveness of chemical cross-linking modification of P84 copolyimide membranes using diamine compounds for PV dehydration has been investigated and the scheme to enhance separation performance of asymmetric polyimide membranes was developed. Two diamine cross-linking agents, *p*-xylene diamine (XDA) and ethylenediamine (EDA), were used in this study for both dense and asymmetric P84 membranes. Experimental results suggest that the cross-linking reaction induced by EDA is much faster than that by XDA because the former has a smaller and linear structure than that of the latter. However, membranes cross-linked by *p*-XDA are thermally more stable than those by EDA. Membranes modified by *p*-XDA or EDA have increased hydrophilicity. An increase in the degree of crosslinking reaction initially results in an increase in separation factor with the compensation of lower permeation rate for PV dehydration of 2-propanol. However, a further increase in the degree of cross-linking reaction may swell up the polymeric chains because of the hydrophilic nature of these diamine compounds, thus resulting in low separation performance. Post treatment after cross-linking reaction can significantly enhance as well as tailor membrane performance because of the formation of charge transfer complexes and the enhanced degree of cross-linking reaction. A low-temperature heat treatment developed PV membranes with high permeation rate and medium separation factor for H_2O/IPA selectivity, whereas a high-temperature heat treatment produced membranes with high separation factor with medium permeation [65].

Four cross-linked PVA/trimesoyl chloride (TMC) membranes with different degrees of cross-linking were prepared by applying TMC/ hexane to the surfaces of dried PVA membranes. FTIR-ATR spectroscopy revealed that PVA-TMC membranes had asymmetric molecular structures. PV properties were studied through water permeation at different temperatures and dehydration of 2-propanol/water mixtures at different temperatures and with different feed water fractions. Results showed that PVA-3TMC had the best overall PV properties among the four PVA-TMC membranes [66].

Chitosan/poly(tetrafluoroethylene) (Chito/PTFE) composite membranes were prepared from casting a γ-(glycidyloxypropyl) trimethoxy silane (GPTMS)-containing chitosan solution on poly (styrene sulfuric acid) grafted expended poly(tetrafluoroethylene) film surface. The adhesion between the chitosan skin layer and the PTFE substrate was pretty good to warrant the high performance of Chito/PTFE composite membranes using in PV dehydration processes on 2-propanol. The Chito/PTFE membrane exhibited a permeation rate of 1,730 g/ (m^2 h) and a separation factor of 775 for H_2O/IPA selectivity of 775 at 70°C on for an aqueous solution of 70 wt % IPA [67].

Chung et al. [68] have developed BTDA-TDI/MDI (P84)/polyether sulfone (PES) dual-layer hollow fibers for pervaporation dehydration of isopropanol (IPA). The effects of spinning conditions, e.g., air gap distances, outer- and inner-layer dope flow rates on membrane formation, morphology and pervaporation performance have been investigated. Compared to the wet-spun dual-layer hollow fibers, macro voids in the outer-layer are significantly suppressed in the dry-jet wet-spun fibers. It is found that the intrusion of non-solvent from the outer-layer may propagate the formation of finger-like macro voids into the inner-layer. Because of different phase-inversion rates in the dual-layers, the effects of elongational stresses in the air gap region on the separation performance of dual-layer hollow fibers are much more complicated than that of single-layer hollow fibers.

Both thermal treatment at elevated temperatures and chemical cross-linking modification by *p*-xylene diamine was investigated. Heat treatment at 200°C increases separation factor with reduced flux. However, unlike single-layer P84 hollow fibers, further increased heat treatment temperature do not enhance separation property of P84/PES dual-layer hollow fibers because of the enhanced sub-layer resistance via the densification of PES layer. The performance enhancement by *p*-xylene diamine cross-linking shows very promising for the P84/PES dual-layer hollow fiber because *p*-xylene diamine induces cross-linking reactions only in the P84 outer-layer. The best separation factor/selectivity is reached at 2h cross-linking time. This study demonstrates that a significant material cost saving can be achieved without sacrificing separation performance by choosing the dual-layer hollow fiber approach with the aid of *p*-xylene diamine cross-linking modification.

Using a solution technique, novel polymeric membranes were prepared by incorporating water-soluble blocked diisocyanate into Chito. The membrane containing 40 mass % of blocked diisocyanate showed the highest separation for H2O/IPA selectivity of 5918 with a permeation rate of 2.20×10^{-2} kg/ (m^2 h) at 30°C for 5 mass % of water. The total permeation rate and permeation rate of water were found to be overlapping particularly for higher cross-linked membranes, suggesting that the membranes developed with higher amount of blocked diisocyanate could be used effectively to break the azeotropic point of water/2-propanol mixture, so as to remove a small amount of water from 2-propanol [69].

Chitosan and hydroxypropyl cellulose (Chito/HPC) blend membranes prepared by solution casting method, followed by cross-linking with urea-formaldehyde-sulfuric acid mixture, were tested for the PV dehydration of 2-propanol. Blend membrane containing 20 wt % HPC (Chito/HPC-20) gave a high selectivity of 11,241 for 10 wt % water containing feed mixture. Comparatively, a low selectivity of 488 was observed for plain cross-linked Chito membrane for the same feed mixture; this value further decreased with increasing water composition of the feed mixture. The permeation rate of the plain Chito membrane increased from 0.074 to 0.246 kg/ (m^2 h) over the feed water compositions of 10-30 wt %. For Chito/HPC-20 membrane, permeation rates increased from 0.132 to 0.316 kg/ (m^2 h). The observed permeation rate was higher for Chito/HPC blend membranes, but lower for the plain cross-linked CS membrane; nevertheless, the H_2O/IPA selectivity was higher for Chito/HPC-10 and Chito/HPC-20 blend membranes. For Chito/HPC-40 membrane, the permeation rate was increased from 0.226 to 0.391 kg/ (m^2 h), but H_2O/IPA selectivity decreased from 453 to 80 over the composition range of 10-30 wt % water in the feed. With increasing temperature, flux increased considerably, but the H_2O/IPA selectivity decreased [70].

Mixed matrix membranes of PVA loaded with phosphomolybdic heteropoly acid (HPA) and cross-linked with GA were prepared by the solution casting technique. At high content (i.e., 7 wt % with respect to weight of PVA) of HPA, the mixed matrix membranes could extract water efficiently on the permeate side with a selectivity of 90,000 and a permeation rate of 0.032kg/ (m^2 h) for 10 wt % of water containing feed mixture. Permeation rate of the mixed matrix membranes decreased with increasing concentrations of HPA; however, a significant improvement in PV performance was observed for HPA-loaded membranes than the pristine PVA [71].

CM-Chito/polysulfone (PS) hollow-fiber composite membranes were prepared through GA as the cross-linking agent and PS hollow-fiber ultrafiltration membrane as the support. The permeation and separation characteristics for dehydration of 2-propanol were investigated by the PV method. The cross-linked CM-Chito/PS hollow-fiber composite membranes had high H_2O/IPA selectivity and promising permeability. The permeation rate and separation factor for H_2O/IPA selectivity were 38.6 g/ (m^2 h) and 3238.5, respectively, using 87.5 wt % of 2-prapanol concentration [72].

Matrimid (R) polyimide asymmetric hollow fibers have been fabricated and applied for PV dehydration of IPA. The effectiveness of thermal annealing at high temperatures and/or chemical crosslinking using 1,3-propane diamine (PDA) on the separation property of these fibers has been investigated. An increase in the degree of cross-linking results in an increase in separation factor for H_2O/IPA selectivity and a decrease in the permeation rate. Thermal annealing alone has failed to improve hollow fiber performance due to the cracks caused by inhomogeneous shrinkage in the heating process. Nevertheless, appropriate application of thermal annealing as a pretreatment for cross-linking can produce fibers with the optimal performance. The pristine hollow fibers had a permeation rate and a separation factor for H_2O/IPA selectivity of 6.2 kg/ (m^2 h) and 7.9, respectively; with this thermal treatment followed by cross-linking, the permeation rate and separation factor changed to 1.8 kg/(m^2h) and 132, respectively.

The formation of charge transfer complexes within the polymer matrix during heat treatment not only assists polymeric chain packing and rigidification but also facilitates more efficient PDA cross-linking. Apparently, PDA molecules could also seal the non-selective cracks (defects). These results indicate the combined thermal and chemical modification possibly is an effective method independent of the initial status of the hollow fiber (e.g., defective or defective free) in revitalizing and enhancing the membrane performance. XRD characterization confirmed a tighter polymer networking in hollow fibers with the cross-linking modification. Comparison between the dehydration of different alcohols revealed that a better separation performance could be obtained for alcohols having a larger molecular cross-section [73].

PV membrane for the separation of diacetone alcohol/water mixtures using commercially available membranes for organic enrichment and dehydration was evaluated. Empirical correlations for the effect of the process parameters of feed concentration, feed temperature, permeate-side pressure, and scale-up were developed. The solvent/water mixture was successfully separated with a PVA based Sulzer PERVAP 2210 dehydration membrane. Various dehydration membranes were evaluated and a comparison of the permeation and separation factor was made. The membrane performance in separating acetone/water mixtures was also studied. An overall model to predict the membrane area needed for a scale-up was developed based on the results [74].

Dense PIC membranes of anionic NaAlg and cationic PEI were prepared and cross-linked with GA for dehydration of alcohol/water mixtures by PV. PV dehydration characteristics of the membranes were determined as a function of PEI content, cross-linking time as well as feed water composition. Transport parameters such as absorption, diffusion and permeability of water and alcohols through the membranes were determined. Among the four different membrane compositions, the PIC containing 40% PEI was found to yield optimum separation data in terms of membrane stability, selectivity and permeability. On the other hand, 10% PEI-containing membranes gave the highest selectivity with the lowest permeation rate at ambient temperature, but the membranes were not sufficiently stable [75].

PIC membranes were prepared by the complexation of protonated chitosan with NaAlg doped on a porous, polysulfone-supporting membrane. The PV characteristics of the membranes were investigated with various alcohol/ water mixtures. The PIC membranes had an excellent PV performance in most aqueous alcohol solutions and that the selectivity and permeability of the membranes depended on the molecular size, polarity, and hydrophilicity of the permeant alcohols. However, the aqueous methanol solutions showed a permeation behavior different from that of the other alcohol solutions. Methanol permeated the prepared PIC membranes more easily than water even though water molecules have stronger polarity and are smaller than methanol molecules [76].

Mordenite (Mo)-filled Chito/PAA PIC membranes were prepared by incorporating mordenite into Chito and PAA blending solution for PV separation of ethylene glycol (EG) aqueous solution. The effects of mordenite content on absorption, diffusion and PV performance of the membranes were evaluated. The permeation rate decreased while the separation factor first increased and then decreased with increasing the mordenite content. The M04-PIC60/40 containing 4 wt % mordenite showed the highest separation factor for H_2O/EG selectivity of 258 with a permeation rate of 165 g/ (m^2 h) for 80 wt % EG in feed at 70°C. High H_2O/EG selectivity was due to the diffusivity of water molecule through *PIC membranes* [77].

Cross-linked Chito for the separation of ethylene glycol /water mixtures membranes were prepared by using phosphoric acid in alcohol baths. The cross-linked membranes were subjected to absorption studies to evaluate the extent of interaction and degree of swelling in pure as well as binary mixtures of the two liquids. The membrane had a good potential for breaking the boiling mixture of ethylene glycol/water since a moderately good selectivity of 234 was obtained at a reasonable permeation rate of 0.37 kg/ (m^2 h). The separation factor for H_2O/EG selectivity was improved with decreasing feed water concentration whereas permeation rate decreased correspondingly. Increasing the membrane thickness decreased the permeation rate but had a less profound effect on the separation factor. Higher permeate pressure caused a reduction in permeation rate and an increase in selectivity [78].

Dehydration of aqueous ethylene glycol solution by PV was studied by surface cross-linked PVA membrane using GA. When the feed mixture was an aqueous solution of 80wt % ethylene glycol, the permeation rate and separation factor for H_2O/EG selectivity at 70°C were 211 g/ (m^2 h) and 933, respectively. The remarkable dependence of water and ethylene glycol concentration in permeate side as well as their activity coefficients within the membrane on feed concentration indicate that a strong coupling effect existed between water and ethylene glycol, which effectively inhibits the permeation of ethylene glycol and thus considerably enhances the H_2O/EG selectivity. With feed temperature increasing, the permeation rate increases but separation factor for H_2O/EG selectivity decreased significantly due to the difference of activation energies between water and ethylene glycol. With feed flow rate increasing, both the permeation rate and separation factor for H_2O/EG selectivity increased correspondingly [77].

Chung et al. [78] have synthesized nano-sized zeolitic imidazolate frameworks (ZIF-8) and prepared polybenzimidazole (PBI)/ZIF-8 mixed matrix membranes (MMMs) with uniform morphology comprising ZIF as high as 58 wt % by directly mixing the as-synthesized ZIF-8 suspension into the polymer dope for pervaporation dehydration of ethanol, isopropanol (IPA) and butanol. Pervaporation test data show that the water permeability of PBI/ZIF-8 (1:1) MMMs is about one order higher than the original PBI membrane (14,000–22,000 vs. 1200–2300 Barrer). The apparent activation energy calculated from the Arrhenius equation indicates that there is a 29.2% reduction of energy barrier for penetrant transports for PBI membranes consisting of 33.7 wt % ZIF-8 nano-particles. As a result, the 33.7 wt % ZIF-8 in PBI membrane increases water permeability four times without much decrease in selectivity for water. The 58.7 wt % ZIF-8 in PBI membrane has a very high water permeability but a relatively low selectivity. Furthermore, the MMMs also show effectiveness in suppressing ethanol-induced swelling on the polymeric matrix owing to its inorganic properties. Vapor absorption studies confirm that the enhancement of pervaporation performance is primarily attributed to the significant improvement in diffusivity selectivity. With high thermal stability, chemical resistance and excellent compatibility with PBI, ZIF-8 nano-particles are promising fillers for enhancing the overall separation performance of low permeability but high performance materials like PBI.

Pervaporation membranes were made by solution blending of polyvinyl alcohol (PVA) and sodium alginate (SA). Accordingly, five different blends with PVA:SA weight ratio of 75:25, 50:50, 25:75, 20:80 and 10:90 designated as PS1, PS2, PS3, PS4 and PS5, respectively, were prepared. Each of these blends was crosslinked with 2, 4 and 6 wt % glutaraldehyde and the resulting fifteen (5×3) membranes were used for pervaporative separation of 90 wt % dioxane in water. The membranes made from PS4 and PS5 were not stable during pervaporation experiments. Among the stable membranes PS3 membrane crosslinked with 2 wt % glutaraldehyde showed the best results for flux and selectivity. Thus, it was filled with nano size sodium montmorillonite filler and used for separation of dioxane–water mixtures over the entire concentration range of 80–99.5 wt % dioxane in water. The membranes were also characterized by mechanical properties, FTIR, SEM, DTA–TGA and XRD [79].

Two novel poly(vinyl alcohol) (PVA)/organosilica hybrid membranes are developed via the sol–gel process, with phenyltriethoxysilane (PTES) and diethoxydiphenylsilane (DEDPS) as crosslinkers and organosilica precursors, and employed for ethanol dehydration via pervaporation [80]. Effects of organosilane type on the membrane morphology, physico-chemical properties and pervaporation performance of the as-fabricated membranes are studied. In addition, the effects of the organo-silane loading and the thermal treatment temperature on the physicochemical properties and pervaporation performance of the membrane are explored systematic-ally. Various characterization techniques (FT-IR, TGA, SEM, etc.) are employed to illuminate physicochemical changes of membranes. The optimal pervaporation performance of PVA–PTES hybrid membrane with 0.4 mmol/g PTES loading and thermally treated at 80°C has a flux of 145 g/(m^2 h) and a separation factor of 1026 for the pervaporation dehydration of 85% ethanol aqueous solution at 40°C.

Acetic acid dehydration is significant in chemical industry. Pervaporation has attracted increasing attention due to its low energy consumption and environmentally friendly process. However, there is still lack of efficient membrane materials for the pervaporation separation of acetic acid/water mixtures. Therefore, developing new material to prepare pervaporation membrane is currently the main task. In this study, an acid stable Zr-MOF NH_2-UiO-66 was synthesized and incorporated into poly(ethyleneimine) (PEI) to form mixed matrix membranes (MMMs) for separating acetic acid/water mixtures. The NH_2-UiO-66/PEI MMMs were deposited on the surface of NaA zeolite tubular substrate to form composite membranes using dip-coating method. The morphologies and structures of the particles and composite membranes were characterized by SEM, EDX, FTIR and contact angle. The effects of membrane preparation conditions on the separation performance were investigated. The results indicated that the NH_2-UiO-66/PEI composite membranes showed good acetic acid dehydration behavior, because of the high porosity and hydrophilicity of the particles. Moreover, the particles had good compatibility with polymer and strong combination with substrate. Therefore, this study may provide a new material and facile strategy for preparing composite membrane in the separation of acetic acid/water mixtures [81].

3.2. Alcohol Selective Membranes for Water/Alcohol Mixtures

Cross-linked PVA composite membranes have been used in commercial PV plants for dehydration of ethanol beyond the azeotrope. However aqueous ethanol solutions that can be produced by bio-fermentation are dilute (about 10 wt %). Therefore, if ethanol/water (EtOH/H_2O) selective membranes with high efficiency can be prepared, the distillation membranes is difficult can be attributed to the fact that ethanol has a larger molecular size water and must be preferentially permeated through the membrane. In fact, permeation and separation in a PV process through dense membranes is based on the solution-process in the first stage to obtain an azeotrope can be replaced which is very advantageous for reduction of energy cost. There are fewer reports on EtOH/H_2O selective membranes compared with those of H_2O/EtOH selective membranes. One reason why the development of efficient high-performance EtOH/H_2O selective diffusion mechanism [82]. Therefore, when it is required that ethanol molecules with larger molecular size preferentially permeate from an aqueous ethanol solution, it cannot be expected to be separated by the diffusion process. Consequently, only a difference of solubility selectivity in the solution process in which both ethanol and water components are dissolved can contribute to the separation.

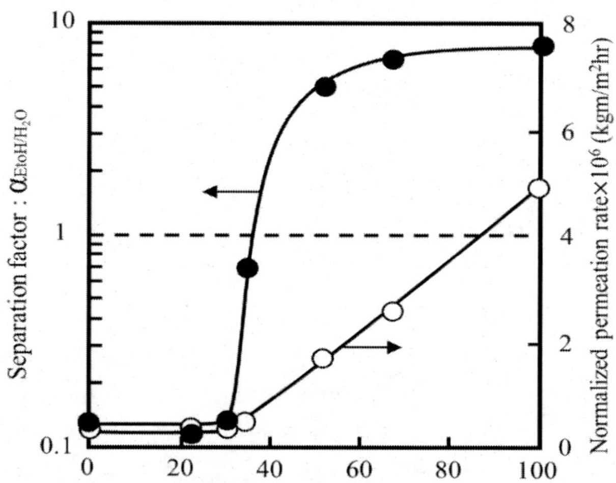

Figure 6. Effect of the DMS content on the separation factor (●) and the normalized permeation rate (○) through the MMA-g-DMS membranes by pervaporation. Feed solution: aqueous solution of 10 wt % ethanol (40°C). Dashed line represents the feed composition.

Figure 6 shows the ethanol concentration in the permeate through a poly (dimethyl siloxane) (PDMS) membrane during PV and that absorbed into a PDMS membrane. These results support the hypothesis that the difference in the solubility of the permeants contributes to the EtOH/H_2O selectivity. PDMS membranes show high EtOH/H_2O selectivity, but their mechanical strength is weak, and it is difficult to prepare thin membranes from PDMS. In order to obtain both EtOH/H_2O selectivity and mechanical strength, graft copolymers composed of PDMS macromonomer and vinyl monomers were synthesized.

Graft copolymer membranes, which were either ethanol- or water-selective, were prepared by copolymerization of an oligo-dimethyl siloxane (DMS) monomer with methyl methacrylate (MMA) [83, 84]. Two glass transition temperatures (T_g) were observed at about 120°C and –127°C in the graft copolymer membranes. Transmission electron micrograph (TEM) demonstrated that the PMMA-*g*-PDMS membranes showed microphase-separated structures. When an aqueous solution of 10 wt % ethanol was permeated through the PMMA-*g*-PDMS membranes by PV, the ethanol concentration in the permeate and the permeation rate increased drastically with the DMS content in the copolymer. In particular, at a DMS content of less than 40 mol %, water permeates preferentially from an aqueous solution of 10wt % ethanol, whereas membranes with more than about 40 mol % of DMS are EtOH/H_2O selective, as shown in Figure 8. The change in the EtOH/H_2O selectivity of the PMMA-*g*-PDMS membranes can be explained by a microphase-separated polymer structure using Maxwell's model and a combined model consisting of both parallel and series expressions. Furthermore, image processing of TEMs allowed the determination of the percolation transition of the PDMS phase at a DMS content of about 40 mol %. These results suggest that the continuity of the PDMS phases in the microphase-separated PMMA-*g*-PDMS membranes directly affects their EtOH/H_2O selectivity for aqueous ethanol solutions.

The EtOH/H_2O selectivity of block copolymer membranes consisting of ethanol-selective PDMS and water-selective PMMA was compared to the EtOH/H_2O selectivity of graft copolymer membranes for the separation of an aqueous ethanol solution. With increasing DMS content, the block copolymer membranes changed from H_2O/EtOH selective to EtOH/H_2O selective at a DMS content of 55 mol %. The graft copolymer membranes showed a dramatic change in their EtOH/H_2O selectivity at a DMS content of 40 mol %. TEMs demonstrated that both membranes had a distinct microphase-separated structure consisting of PDMS and PMMA phases, and that the morphology was quite different between the block and graft copolymer membranes. The morphological changes in these membranes were investigated by image processing of micrographs and analysis using a combined model consisting of both parallel and series models. These investigations revealed that the percolation transition of the PDMS phase in the block and graft copolymer membranes takes place at a DMS content of about 55 and 40 mol %, respectively. This suggests that the continuity of the PDMS phase in these microphase-separated membranes strongly influences their ethanol selectivity [85].

The effects of annealing on selectivity during PV was also investigated for these block and graft copolymer membranes. The EtOH/H_2O selectivity of the block copolymer membranes was strongly influenced by annealing, but that of the graft copolymer membranes was essentially not affected. The original block copolymer membranes changed from being water- to ethanol-selective at a DMS content of 55 mol %, but the annealed block copolymer membranes changed at a DMS content of 37 mol %. TEMs demonstrated that the annealing of block copolymer membranes with a DMS content between 37 and 55 mol % resulted in dramatic changes in their morphology. However, annealing of the graft copolymer membranes had very little effect on their microphase- separated morphology, which was quite different from the morphology of the block copolymer membranes. Again, an analysis using a combined model consisting of parallel and series models revealed that a continuous PDMS phase in the direction of the membrane thickness was readily formed by annealing of the block copolymer membranes. As a result, the continuity of the PDMS phase in the microphase-separated structure governed the ethanol selectivity of these membranes for an aqueous ethanol feed solution [86].

It is well known that poly[1-(trimethylsilyl)-1-propyne] (PTMSP) membranes show high EtOH/H_2O selectivity [87, 88]. To enhance the EtOH/H_2O selectivity of PTMSP membranes, surface- modified PTMSP membranes were prepared by adding a small amount of a polymer additive, a graft copolymer (PFA-*g*-PDMS) consisting of poly(Fluor acrylate) (PFA) and PDMS, in the casting solution of PTMSP. Modified PTMSP membranes were cast on glass plates and the contact angles of water on the membrane surfaces exposed to the air- side and the glass- side, respectively, were measured. The contact angle for water on surface- modified PTMSP membranes was significantly different on the air- side versus that on the glass side; the contact angles on the air- side were more hydrophobic. Furthermore, the contact angle for water increased in hydrophobicity with additional amounts of PFA-*g*-PDMS. The high hydrophobicity of the membrane surface on the air-side and the increase in hydrophobicity with additional amounts of polymer additive were also confirmed by X- ray photoelectron spectroscopy. The permeation rate for an aqueous solution of 10 wt % ethanol in PV experiments using surface- modified PTMSP membranes decreased slightly. However, the EtOH/H_2O selectivity increased considerably with increasing amounts of PFA-*g*-PDMS [89, 90].

An integrated fermentation and membrane-based recovery PV process has certain economic advantages in continuous conversion of biomass into alcohols. New PV data obtained for PTMSP samples synthesized in various conditions were reported. Three different catalytic systems, $TaCl_5$/*n*-BuLi, TaC15/Al(*i*-Bu) (3), and $NbCl_5$ were used for synthesis of the polymers. The catalytic system has a significant influence over the properties of membranes made from PTMSP. Although a combination of a high permeation rate and a high separation factor for EtOH/H_2O selectivity (not less than 15) was provided by all PTMSP samples, the PTMSP samples synthesized with $TaCl_5$/*n*-BuLi showed significant deterioration of membrane properties when acetic acid was present in the feed. In contrast, the PTMSP samples synthesized with $TaCl_5$/Al(*i*-Bu) (3) or $NbCl_5$ showed stable performance in the presence of acetic acid. When using a multicomponent mixture of organics and water, the permeation of different organic components results in lower separation factor for both ethanol and butanol. These data are consistent with Nano porous morphology of PTMSP. It was demonstrated that pervaporative removal of ethanol improved the overall performance of the fermentation process [91].

PV can be utilized for recovery of ethanol from fermentation broths. A hybrid fermentation-PV-distillation continuous process was considered for the production of 30, 000 ld^{-1} of 95% ethanol. Available data of composite hollow fiber poly dimethyl siloxane membrane of PV module was used to develop the model equations. Direct production cost of ethanol was estimated and used for optimization of ethanol concentration in fermenter and retentate. Minimum direct production cost of 0.2 $/L of ethanol was found at 55 kg/m^3 of ethanol mass concentration in fermenter and 50 kg/m^3 in retentate. A sensitivity analysis of the PV performance parameters; permeation rate, separation factor and membrane cost was also carried out [92].

A novel inorganic/organic composite membrane was prepared through introducing carbon black into PDMS membrane. Carbon black was treated with various measures, including extraction, methylolation, and high-temperature calcinations in order to tailor the nature of the surface. The effects of surface treatment, carbon black loading, particle size, and temperature on the PV performance in the extraction of ethanol from ethanol/water mixtures were explored. In certain ranges of composition, the permeation rate was remarkably increased without reducing the EtOH/H_2O selectivity [93].

Pervaporation is a potential process for recovering bioethanol produced from biomass fermentation. Fermentation broths contain ethanol, water, and a variety of other compounds, often including carboxylic acids. The effects of acetic acid on long-term pervaporation of aqueous ethanol mixtures through high-silica ZSM-5 zeolite-filled polydimethylsiloxane (PDMS; silicone rubber) membranes were investigated. Acetic acid was shown to reduce the ethanol removal effectiveness of these membranes. Initially after acetic acid addition, ethanol and water fluxes decreased due to acetic acid competing with ethanol and water for absorption sites in the membrane. Longer-term exposure to acetic acid resulted in an irreversible, steady decline in ethanol/water separation factor because of declining ethanol flux. Increasing feed pH to above the dissociation constant (pK_a) of acetic acid diminished the longer-term decline in ethanol flux and essentially eliminated the effect of competitive absorption. Measurements of absorption competition between ethanol, water, and either acetic acid or succinic acid on the zeolite particles suggested that other carboxylic acids would have qualitatively similar short-term effects on membrane performance as those observed for acetic acid.

Pervaporation is a potential process for recovering bioethanol produced from biomass fermentation. Fermentation broths contain ethanol, water, and a variety of other compounds, often including carboxylic acids. The effects of acetic acid on long-term pervaporation of aqueous ethanol mixtures through high-silica ZSM-5 zeolite-filled polydimethylsiloxane (PDMS; silicone rubber) membranes were investigated. Acetic acid was shown to reduce the ethanol removal effectiveness of these membranes. Initially after acetic acid addition, ethanol and water fluxes decreased due to acetic acid competing with ethanol and water for absorption sites in the membrane. Longer-term exposure to acetic acid resulted in an irreversible, steady decline in ethanol/water separation factor because of declining ethanol flux. Increasing feed pH to above the dissociation constant (pK_a) of acetic acid diminished the longer-term decline in ethanol flux and essentially eliminated the effect of competitive absorption. Measurements of absorption competition between ethanol, water, and either acetic acid or succinic acid on the zeolite particles suggested that other carboxylic acids would have qualitatively similar short-term effects on membrane performance as those observed for acetic acid [94].

The separation of ethanol/water could be cost-competitive using PV in the production of renewable biomass ethanol, but the performance of the modified or unmodified polymeric membranes still is not satisfactory. For the purpose of improving the PV performance of polymeric membranes, especially for permeation rate, PDMS was deposited uniformly on the surface of tubular asymmetric ZrO_2/Al_2O_3 porous ceramic supports. The thickness of the PDMS layer formed atop the ZrO_2 layer was on the order of 5-10 mμm. In the PV experiment of ethanol/water mixtures, as the ethanol concentration increased, the total permeation rate increased but the $EtOH/H_2O$ selectivity decreased. At the same time, with increasing operating temperature, the total permeation rate of the composite membranes increased whereas the $EtOH/H_2O$ selectivity decreased. It was observed that the PDMS/ceramic composite membrane showed a great total permeation rate of 19.5 kg/(m^2 h) and separation factor for $EtOH/H_2O$

selectivity of 5.7 at a feed temperature of 70°C under a pressure of 460 Pa in an ethanol concentration of 4.3 wt %. The total permeation rate of the PDMS/ceramic composite membranes were superior to other reported PDMS membranes [95].

Studies were carried out to improve the effectiveness and cost-effectiveness of a semi-continuous ethanol fermentation of lactose mash combined with a PV module [96]. During a 20-day fermentation/PV with an immobilized biocatalyst, the fermentation of the lactose mash (12%) averaged 4.56% m/v ethanol. In the circulation PV module, ethanol from about 2,000 g portions of the fermented mash was removed to below 0.7% m/v in 10-12 h. The effectiveness of ethanol separation ranged between 88 and 95% and was determined by the ethanol concentration in the fermented mash. The productivity of 15.6% m/v ethanol obtained in such a system was about 530 g/day. The PDMS-PAN-PVA membrane applied to the PV system proved to be highly selective towards ethanol, separation factor for EtOH/H_2O selectivity was > 8, and permeation rate was 2,600-3,500 g/(m^2 h).

A study was conducted to separate butanol from an aqueous solution using pervaporation. A specially designed and manufactured cell was used to separate the butanol from butanol/water solutions of different butanol concentrations (6-8-11-16-20-50) g/l. A 250 cm^3 butanol mixture at 33°C was used to feed the cell, while the pressure of permeation side was about ~0 bar. Results revealed that butanol concentration changes non-linearly during the first 3 h, and then proceeds linearly. The percentage of butanol removal increases with increasing feed concentration. The permeability of the used membrane was determined experimentally. A resistance in series model was used to simulate the pervaporation step. The butanol concentration in the feed during the pervaporation step was predicted by using the developed model. There is a fair agreement between butanol concentration in feeding tank of pervaporation cell both experimentally and predicted from the developed model [97].

PV experiments were conducted to recover 1-butanol (BuOH) from model pharmaceutical aqueous waste using a surface modified poly (vinylidene fluoride) (PVF_2) membrane. The surface modification of the membrane was made using silicone grease as an ultra-thin layer on the surface to improve the PV performance of the membrane. The effect of operating variables such as feed composition, feed temperature and feed flow rate on permeation rates, separation factor and pervaporative separation index were studied in order to optimize the operating variables. The experimental results showed that surface modified PVF_2 membrane was BuOH/H_2O selective especially for low feed compositions. The separation factor for BuOH/H_2O selectivity of 6.4 and total permeation rate of 4.126 k/ (m^2 h) were obtained at a feed composition of 7.5 wt %, feed temperature of 50°C, feed flow rate of 600 mL/min and permeate pressure of 50 mmHg. The total permeation rate of the surface modified membrane increased with increasing feed composition, feed temperature, feed flow rate of the mixture whereas the separation factor for BuOH/H_2O selectivity was a reversed order except for flow rate. The influence of operating variables such as feed composition and temperature on partial permeation rate and permeate composition was modelled based on Fick's first law to understand the process behavior and it will be very useful for design purpose. These models will be used to predict the required membrane surface area for recovery of 1-butanol for the range of experimental feed compositions [98].

The objective of this work was to study the effect of silane coupling agent on the performance of mixed matrix membranes (MMMs) and to develop a mathematical model to analyze the performance of the MMMs. The silicalite-1 modified by various alkoxysilanes was incorporated into polydimethylsiloxane (PDMS) to prepare dense MMMs. The modified silicalite-1 and corresponding MMMs were characterized by FT-IR, CA, TGA, DSC and SEM. These results confirmed that the silicalite-1 was successfully modified by various alkoxysilanes. The effects of alkoxysilane chain length and chemical structure on the pervaporation (PV) performance of MMMs were discussed in detail. All the silane modification did not change the framework of silicalite-1. Better dispersion of silicalite-1 in PDMS was attained after modification and all the modified membranes could eliminate the nonselective voids inside the membranes. Moreover, the silane coupling agent had a contradictory effect on the PV performance of the membranes. On the one hand, the silane coupling agent improved the compatibility between silicalite-1 and PDMS. On the other hand, the silane coupling agent introduced a silane phase around the silicalite-1, which depressed the selectivity of the pristine silicalite-1. Silicalite-1 grafted with vinyl group showed the best compatibility with PDMS, and the resulting MMMs had the highest separation factor to ethanol because strong chemical bonds between vinyl-silicalite-1 and PDMS were formed with the thinnest silane film. For the separation of ethanol from a dilute solution, a separation factor of 34.3 was obtained with the vinyl-MMMs, which was 49% higher than that of unmodified membrane. As the silicalite-1 content increased, the separation factor increased. Furthermore, a mathematical model was proposed to predict the performance of MMMs. And both the permeability and selectivity obtained by the proposed model were in good agreement with the experiment data [99].

Zeolitic imidazolate framework (ZIF-7)/polydimethylsiloxane (PDMS) mixed matrix membranes (MMMs) with uniformly dispersed ZIF-7 were fabricated by Li et al. [100]. The homogeneous dispersion of ZIF-7 nanoparticles in PDMS matrix and better interface compatibility were characterized by SEM and EDS. PV performance reveals that the MMMs display increased total flux and the highest separation factor occurs at 20 wt % ZIF-7 loading, which were 1689 g/(m^2 h) and 66 respectively as compared to 1080 g/(m^2 h) and 51 for pure PDMS membrane when separating 1 wt % butanol aqueous solution at 60°C. The total flux reaches 3496 g/ (m^2 h) with 80 wt % butanol in permeate for the enrichment of 5 wt % butanol solution at 60°C. The enhanced flux may result from enlarged free volume in the polymer matrix caused by the incorporation of ZIF-7 nanoparticles. Moreover, the super-hydrophobic ZIF-7 pore channels and butanol preferential permeation may contribute to the improvement of separation factor. The result also shows that there is a 21.1% reduction of energy barrier for penetrant transportation for 20 w t% ZIF-7 filled PDMS membrane as compared to pristine PDMS membrane. A long period of pervaporation experiment was also carried out to investigate the structure stability of the ZIF-7/PDMS membrane, and the MMM remains intact during the 240 h continuous operation, which is the key to industrial application for membrane-based separation process. All above indicated that ZIF-7 is a good filler to fabricate butanol permselective pervaporation membranes for butanol separation and enrichment.

Separation of ethanol/water mixtures through classical separation methods, like distillation, are very energy consuming. Pervaporation is an alternative membrane separation process with much lower energy demand, but still lacking high performance membranes with sufficient flux-selectivity properties. In this study, polydimethylsiloxane (PDMS) based mixed matrix membranes (MMMs) were developed and their pervaporation performance investigated. The fillers consist of hollow spheres (HS) covered with a shell of silicalite-1 crystals. These HS were characterized by scanning electron microscopy (SEM), transmission electron microscopy (TEM), nitrogen physisorption and X-ray diffraction (XRD). The spheres were approximately 1 μm in size with a shell thickness of 30 nm. Nitrogen physisorption revealed the micro- and mesoporous nature of the spherical shells, with a BET surface area of over 800 m^2/g. The hollow silicalite spheres were then uniformly distributed in the PDMS membrane to increase the membrane permeability since the hollow core of the HS allows very fast flow of the permeating compound. Furthermore, the zeolitic shell improves the ethanol selectivity through its specific pore structure and hydrophobicity, while additional crosslinking of the HS with the PDMS matrix further increases the selectivity of the polymer matrix, thus reaching a separation factor and flux value of 16 and 3.8×10^{-6} g m/ (m^2 s) respectively for a 6% aqueous ethanol solution at 40°C [101].

Membranes that enable the recovery of organic compounds from diluted aqueous solutions are desired for applications such as biobutanol production. The polymer of intrinsic micro porosity PIM-1 shows promise for organophilic separations and here it is incorporated into thin film composite (TFC) membranes in order to increase the flux of permeate. Asymmetric polyvinylidene fluoride (PVDF) supports were prepared with pore sizes at the surface in the size range 25–55 nm and fractional surface porosities in the range 0.38–0.69, as determined by atomic force microscopy (AFM). The addition of phosphoric acid to the PVDF dope solution helped to control the pore size and porosity. Supports were coated with PIM-1 to form TFC membranes with active layer thicknesses in the range 1.0–2.9 μm.

Membranes were tested for the pervaporation of a 1-butanol/water mixture (5 wt %). At 65°C, values of total flux up to 9 kg /(m^2 h) were obtained, with separation factors up to 18.5 and values of pervaporation separation index (PSI) up to 112 kg/(m^2 h) [102].

Sedimentation of silicalite-1 occurs in the fabrication of thin silicalite-1 filled polydimethylsiloxane (PDMS) hybrid composite membranes if the viscosity of membrane solution is low, which makes this preparation challenging. In this work, a new method that use a platinum catalytic agent to assist the pre-polymerization of PDMS polymer to increase the viscosity of the membrane solution was studied [103]. With this method, supported silicalite-1 filled PDMS hybrid composite membranes were fabricated and applied in the pervaporative separation of a 5 wt % dilute ethanol aqueous solution. The effect of the concentration of platinum catalytic agent on the membrane properties was first investigated using CRM, DSC and extraction experiment. Optimum of viscosity of the composite membrane solution was then conducted and a selective layer of as thin as 5 μm thickness was obtained with a flux of 5.52 kg/ (m^2 h) in combination with a separation factor of 15.5 at 50°C. After that the separation performances of different thick membranes, interfacial adhesion properties of hybrid membranes, comparisons with other reported results and membrane stability were investigated. Results showed homemade silicalite-1-PDMS hybrid composite membrane offers relatively high separation performance, indicating a potential industrial application for the separation of ethanol from aqueous solutions.

3.3. Water/Organic Liquid Selective Membranes

Water/organic selective membranes are effective for the dehydration of water/organic mixtures. The dehydrated organic solvents can be usefully as industrial reaction solvents, washing solvents, and analytical solvents.

Polyvinyl alcohol (PVOH) membrane, was modified both physically and chemically by incorporation of inorganic filler, sodium aluminosilicate and chemical crosslinking with maleic acid and glutaraldehyde. The change of morphology and crystallinity of PVOH by this physical and chemical modification was studied by FTIR, DSC, TGA, SEM and XRD. These membranes were evaluated in terms of its potential for dehydration of dioxane by preferential absorption and permeation using pervaporation (PV) technique. These membranes were cast in the laboratory by solution casting from the polymer and other additives. The performance of the unfilled (containing no filler) glutaraldehyde (GA) crosslinked PVOH-1 and maleic acid (MA) crosslinked PVOH-2 membranes were compared with filled (containing aluminosilicate filler) but GA crosslinked PVOH-3 and filled but MA crosslinked PVOH-4 membranes. The filled membranes were found to show higher flux and water selectivity. Among all the four used membranes, the MA crosslinked filled PVOH-4 membrane was found to show best results in terms of both water selectivity and flux [104].

Novel hybrid composite membranes have been prepared by incorporating 5 and 10 wt % of sodium montmorillonite (NaMMT) clay particles into NaAlg and cross-linked with GA. PV separation performance of the hybrid composite membranes were investigated for the dehydration of 2-propanol, 1,4-dioxane (DIOX) and tetrahydrofuran from their aqueous solutions. NaMMT particles could be intercalated in the aqueous polymer solution. The driving force for NaMMT absorption is entropic, which involves at least partial replacement of water of hydration associated with exchangeable cations in the clay galleries. The results of PV experiments demonstrated that the addition of NaMMT clay particles increased the selectivity to water over that of pristine NaAlg membrane. Permeation rates of the hybrid composite membranes were lower than those observed for plain NaAlg membrane [105].

Microporous alumino-phosphate (AlPO4-5) has been employed to prepare novel NaAlg-based composite membranes by solution casting technique and cross-linking with GA. These membranes were tested for the PV dehydration of 2-propanol (12.6 wt % water), 1,4-dioxane (18.1 wt % water), tetrahydrofuran (6.7 wt % water) and ethanol (4 wt % water) from their aqueous mixtures. However, PV dehydration studies at feed composition from 5 to 20 wt % were done for pristine NaAlg and 20 wt % AlPO4-5-loaded composite membranes. The activation parameter values involved in the permeation process were evaluated. The permeation rate and selectivity for a near the azeotropic compositions of the feed mixtures were enhanced with increasing AlPO4-5 content into NaAlg-based matrix. The selectivity to water was higher for water/IPA azeotrope, but the permeation rate was more in case of water/1,4-dioxane and water/tetrahydrofuran azeotropes. The selectivity to water was smaller for water/tetrahydrofuran and water/ethanol azeotropes as compared to water/2-propanol and water/1,4-dioxane azeotropes. Molecular sieving effect due to uniform distribution of microporous molecular sieve particles, and hydrophilic characteristics of the alumino-phosphate molecular sieve in addition to its interaction with the hydrophilic NaAlg matrix were responsible for such an appreciable increase in membrane performance over that of pristine cross-linked NaAlg membrane [106].

The blocked diisocyanate crosslinked chitosan membrane was modified by incorporating different mass % of NaY zeolite. The physio-chemical properties of resulting composite membranes were studied using Fourier transform infrared spectroscopy (FTIR), wide-angle X-ray diffraction (WAXD), thermogravimetric analysis (TGA), differential scanning calorimetry (DSC) and scanning electron microscopy (SEM). The mechanical properties of the membranes were studied using universal testing machine (UTM). After measuring the equilibrium swelling, membranes were subjected to pervaporation for separation of water–isopropanol mixtures. Both flux and selectivity were increased with increasing NaY zeolite content in the membranes.

The membrane containing 40 mass % of NaY zeolites exhibited the highest separation selectivity of 11,241 with a flux of 11.37×10^{-2} kg/ (m^2 h) for 10 mass % of water in the feed. The total flux and flux of water are almost overlapping each other, suggesting that these membranes could be effectively used to break the azeotropic point of water–isopropanol mixture. From the temperature dependent diffusion and permeation values, the Arrhenius activation parameters were estimated. All the composite membranes exhibited lower activation energy compared to crosslinked membrane, indicating that the permeants require less energy during the process because of molecular sieving action attributed to the presence of sodalite and super cages in the framework of NaY zeolite. The Henry's mode of absorption dominates the process, giving an endothermic contribution [107].

The performance of pervaporation dehydration of a crosslinked polybenzoxazine membrane (CR-PBz-M) on isopropanol (IPA) aqueous solutions has been studied [108]. Compared to the feature showing in dry state, CR-PBz-M exhibits a change in micropore size distribution from a single distribution to a bimodal pattern in wet state being swollen with the feeding solutions. The presence of two groups of micropores in different sizes attributes to the high pervaporation dehydration performance of the membrane, in which the small micropores resist to the permeation of IPA molecules so as to contribute to the high membrane selectivity and the large micropores and high micropore volume fraction contribute to the high permeation fluxes. This feature of CR-PBz-M demonstrates an in situ self-promoted characteristic for pervaporation dehydration membranes. A permeation flux of 330 g/ (m^2 h) and a 100% of water in the permeate side has been recorded on pervaporation dehydration on a 70 wt % IPA aqueous solution.

PVA-based nanocomposite membranes were prepared by coprecipitation of different amounts of Fe(II) and Fe(III) taken in an alkaline medium and their PV performances were investigated to dehydrate from aqueous solutions of 10-20 wt % 2-propanol and 1,4-dioxane, and 5-15 wt % tetrahydrofuran. Thin layer membranes were cast on polyester fabric cloths as support layers to improve their PV separation performances for all the three mixtures over that of the pristine cross-linked PVA membrane. In particular, the composite membrane prepared by taking 4.5 wt % of iron oxide showed an improved selectivity with a slight sacrifice in the permeation rate compared to membranes containing lower contents of iron oxide as well as the pristine cross-linked PVA membrane. The permeation rate decreased with increasing content of iron in the PVA matrix, while the selectivity increased systematically [109].

The PV separation of 1,4-dioxane/water mixtures was carried out using cross-linked blend membranes of Chito and nylon 66 (N66). Optimum Chito/N66 ratio was determined as 90/10 (w/w) for the containing water of

4.3 wt % at 40°C [110]. Increasing barrier from 30 to 120 mμm improved separation factors for H_2O/DIOX selectivity from 767 to 1,123, but permeation rates were lowered from 0.118 to 0.028 kg/(m ^{2}h). An azeotrope of 1,4-dioxane/water mixture (1,4- dioxane of 82 wt %) was easily broken with a separation factor for H_2O/DIOX selectivity of 865 and permeation rate of water was 0.089 kg/(m^2 h).

Acrylonitrile (AN) was copolymerized with 2-hydroxyethyl methacrylate (HEMA) at three different copolymer compositions [111] by emulsion polymerization to produce PAN/HEMA copolymer membranes. These membranes were PAN/HEMA-1, PAN/HEMA-2 and PAN/HEMA-3. The PV dehydration of tetrahydrofuran over a concentration range of 0-14 wt % water in the feed through these three copolymer membranes was studied. Among the copolymer membranes, PAN/HEMA-1 membrane exhibited a reasonable permeation rate of water (34.9 g/(m^2 h) with a very high H_2O/THF selectivity of 264, whereas PAN/HEMA-3 membrane showed a higher permeation rate of water of 52 g/(m^2 h) but the H_2O/THF selectivity for highly concentrated tetrahydrofuran (water of 0.56 wt % in the feed) at 30°C was 176.5.

Cross-linked blend Chito/PVA membranes were prepared by casting mixtures of Chito and PVA and a urea formaldehyde/sulfuric acid mixture. Chito was used as the base component in the blend system, whereas PVA concentration was varied from 20 to 60 wt %. Membranes were tested for PV dehydration of 2-propanol and tetrahydrofuran. In close proximity to their azeotropic compositions, the membrane performance was assessed by calculating permeation rate and selectivity. Swelling experiments performed in water/ organic mixtures were used to explain the PV results. The cross-linked blend membrane containing 20 wt % PVA was tested for water of 5 and 10 wt % in the tetrahydrofuran/water mixture and 2-propanol/water mixture. Their separation factors for water selectivity were 4,203 and 17,991, respectively. The permeation rate increased with increasing concentration of water in the feed. The selectivity was the highest for the cross-linked blend membrane with PVA of 20 wt % [112].

Dense polymer membranes were made by mixing aqueous solutions of hydrophilic polymers PVA and PEI for investigating the separation of an azeotrope of tetrahydrofuran/water (94 wt % tetrahydrofuran) by PV. The membranes were found to have good potential for breaking an azeotrope of tetrahydrofuran. An increase in PVA content in the blend caused a decrease in the permeation rate and an increase in selectivity. The blend membrane of PVA/PEI of 5/1 showed the highest separation factor for H_2O/THF selectivity of 181.5 and the permeation rate of 1.28 kg/ (m^2 h) for an azeotropic mixture, respectively [113].

Mixed matrix polymer membranes containing nanosized (21 nm) TiO_2 particles dispersed in NaAlg were prepared by solution casting and cross-linking with GA. These membranes were tested for PV dehydration of tetrahydrofuran and 2-propanol from their aqueous solutions. Plain cross-linked NaAlg membrane could remove up to a containing water of 97 wt % for the feed with high water content. The mixed matrix membranes of NaAlg had infinite selectivity for the dehydration with reasonable permeation rates. Permeation rates of NaAlg-TiO_2 mixed matrix membranes were slightly lower than those of plain the NaAlg membrane [114].

Zeolite K-LTL-loaded NaAlg-mixed matrix membranes were prepared by solution casting and cross-linked with GA. The PV dehydration of 2-propanol, 1,4-dioxane and tetrahydrofuran was tested at 30-70°C as a function of membrane thickness and feed compositions. These membranes showed enhancement in both the permeation rate and water selectivity at azeotropic mixtures. These results were due to the addition of K-LTL particles in NaAlg matrix. The permeation rate and selectivity to water were higher for water/1,4-dioxane azeotrope than those of water/2-propanol and water/ tetrahydrofuran azeotropic mixtures. Molecular sieving effect based on uniform distribution of K-LTL zeolite particles, in NaAlg matrix its hydrophilicity and hydrophilic NaAlg gave high membrane performance than pristine cross-linked NaAlg membrane [115].

Permeation and separation characteristics of *N, N*-dimethylformamide (DMF)/water mixtures were investigated by PV, EV and TDEV using alginate membranes cross-linked with calcium chloride. The permeation rate was found to be inversely proportional to the membrane thickness whereas the separation factor increased with increasing membrane thickness. The permeation rates in EV and TDEV were lower than those in PV, and the highest separation factors were obtained with TDEV. Alginate membranes gave permeation rates of 0.97-1.2 kg/ (m^2 h) and separation factors for H_2O/DMF selectivity of 17-63 depending on the operation conditions and the membrane separation method. The absorption selectivity was dominant factor for the separation of *N, N*-dimethyl formamide/water mixtures [116].

Membranes constructed from pure poly (4-methyl-1-pentene) (TPX) and 4-vinylpylidine (4-VP) modified TPX membranes (TPX/P4-VP) were also prepared for PV. The introduction of a hydrophilic 4-VP monomer into the TPX matrix was done by free radical polymerization to form the TPX/P4-VP membrane. The separation factor for H_2O/CH_3COOH selectivity and permeation rate of the TPX/P4-VP membranes were higher than those of the unmodified TPX membranes for the PV of an aqueous acetic acid solution. A good relationship was obtained between the water concentration in the feed and the permeation rate of water by applying the Michaelis-Menten equation [117].

The PV separation of acetic acid/water mixtures was carried out over the full range of compositions at 30-55°C using PVA membranes modified with PAA. The best condition for the preparation of the membranes was found as PVA/PAA ratio as 75/25 (v/v). PVA/PAA membranes gave separation factors of 34 to 3548 and permeation rates of 0.03 to 0.60 kg/ (m^2 h), depending on the operation temperature and feed mixture composition [118].

Itaconic acid (IA) was grafted onto PVA with cerium (IV) ammonium nitrate as an initiator. IA-*g*-PVA membranes were prepared with a casting method, and permeation and separation characteristics of acetic acid/water mixtures were investigated by PV, EV and TDEV. The permeation rates in EV were lower than those in PV, whereas the separation factors for H_2O/CH_3COOH selectivity in EV were higher. In TDEV, permeation rates decreased and separation factors for H_2O/CH_3COOH selectivity increased with dropping temperature of the membrane surroundings. The highest separation factor for H_2O/CH_3COOH selectivity of 686 was obtained for acetic acid of 90 wt % in the feed in TDEV [119].

The separation of acetic acid/water mixtures was carried out using PV and TDEV. For the separation process, 4-vinyl pyridine was grafted on PVA. Membranes were prepared from the graft-copolymer by casting method and cross-linked by heat treatment. Permeation rates were found to be high in PV whereas separation factors for H_2O/CH_3COOH selectivity were high in TDEV. Membranes gave permeation rates of 0.1-3.0 kg/ (m^2 h) and separation factors of 2.0-61.0 depending on the composition of the feed mixture and the membrane separation method [120].

Aacrylonitrile (AN) and hydroxy ethyl methacrylate (HEMA) were grafted onto poly (vinyl alcohol) (PVA) using cerium (IV) ammonium nitrate as initiator at 30°C. The PVA-*g*-AN/HEMA membranes were prepared by a casting method, and used in the separation of acetic acid/water mixtures by PV. PVA-*g*-AN/HEMA membranes gave separation factors for H_2O/CH_3COOH selectivity of 2.26-14.60 and permeation rates of 0.18-2.07 kg/ (m^2 h). Grafted membranes gave lower permeation rates and greater separation factors for H_2O/CH_3COOH selectivity than PVA membranes [121].

Novel composite membranes were prepared from two hybrid materials of class I and class II. Membranes exhibited a remarkable increase in the degree of swelling with increasing zeolite loading in class II hybrid material. The PV performance of these membranes for the separation of water/acetic acid mixtures was investigated in terms of feed composition and zeolite loading. Both the permeation rate and H_2O/CH_3COOH selectivity increased simultaneously with increasing zeolite content in the membrane matrix. These results were explained based on enhancement of hydrophilicity, selective absorption and the establishment of molecular sieving action. Among the membranes developed, the membrane containing 15 mass % of zeolite exhibited the highest separation factor for H_2O/CH_3COOH selectivity of 2,423 with a permeation rate of 8.35 x 10^{-2} kg/(m^2 h) for 10 mass % of water in the feed at 30°C [122].

A new type of sodium alginate (NaAlg) composite membrane by incorporating cobalt (III) (3- acetylpyridine-*o*-aminobetizoylhydrazone) (Co-APABZ) complex as filler particles in different ratios were developed. Membranes were prepared by solution casting followed by solvent evaporation and cross-linked with GA. NaAlg composite membranes in the presence of Co-APABZ particles preferentially absorbed water molecules to facilitate diffusion of water through the membranes and thus enhance the H_2O/CH_3COOH selectivity in the acetic acid/water mixture. However, the membrane performance was significantly affected by the content of Co-APABZ in the NaAlg matrix and the degree of swelling of the membrane. The membrane with Co-APABZ of 5 wt % in NaAlg matrix resulted in the separation factor for H_2O/CH_3COOH selectivity of 174 and the permeation rate of 0.123 kg/ (m^2 h) for an aqueous solution of 90 wt % acetic acid [123].

Novel hydrophilic polymer membranes based on cross-linked mixtures of poly (allylamine hydrochloride)-PVA are developed. The high selectivity and permeation rate characteristics of these membranes for the dehydration of organic solvents are evaluated using PV technology and are found to be very promising when compared to existing membranes [124].

To align poly(3-hydroxybutyrate) (PHB)-functionalized multi-walled carbon nanotubes (MWCNTs) into a chitosan matrix was tried by Tan et al. [125]. The MWCNTs were first functionalized with PHB and aligned into a membrane filter template through a filtration process. A solution casting technique was then applied to cast the chitosan onto the template to form PHB-MWCNT/chitosan nanocomposite membranes. The functionalized MWCNTs and resulting membranes were characterized using thermogravimetric analysis (TGA), Fourier transform infrared spectroscopy (FTIR), scanning electron microscopy (SEM) and transmission electron microscopy (TEM). The TGA demonstrated that approximately 40 wt % of the PHB was successfully functionalized in the MWCNTs, and TEM showed that the polymer wrapped on the nanotubes surface. The presence of the PHB functional moieties helped to improve the dispersion and compatibility of the MWCNTs in the chitosan matrix. When we applied the PHB-MWCNT/chitosan nanocomposite membranes in the pervaporation process of 1,4-dioxane dehydration, the nanocomposite membrane showed a relatively high permeation flux and selectivity towards water, compared to existing membrane.

The functionalization of multi-walled carbon nanotubes (MWCNTs) with poly (vinyl alcohol) (PVA) to improve the compatibility and dispersion of MWCNTs in a chitosan (CS) matrix is studied by Tan et al. [126]. The resultant PVA–MWCNT/CS nanocomposite membranes were crosslinked with glutaraldehyde. Pervaporation performances of the resultant membranes in dehydration of acetone were evaluated in terms of water permeance and selectivity towards water. The water permeance of the crosslinked nanocomposite membrane increased while the selectivity decreased compared to the crosslinked pure CS membrane. In addition, the selectivity and water permeance of the crosslinked nanocomposite membrane decreased while the water permeation flux improved with increasing feed temperature. Furthermore, crosslinking of the membranes was found to improve the selectivity of the membranes but lower the water permeance. In another approach, PVA functionalized MWCNT was bulk aligned on the poly (vinylidene fluoride) (PVDF) membrane by a simple filtration method and further coated with CS to form a novel three-layer nanocomposite membrane. This membrane showed immense improvement on the water permeance and selectivity. Upon comparison, the three-layer nanocomposite membrane was found to demonstrate the best separation performance among the membranes described here. Furthermore, the three-layer nanocomposite membrane emerged as a potential solution to the trade-off problem often faced by pervaporation membranes.

3.4. Organic Selective Membranes for Organic Liquid/Water Mixtures

Organic liquid/water selective membranes are effective for the removal of organics in water and recovery of organic solvents from water. These membranes can be contributed to the environmental problem and effective use of organic solvents.

The removal and enrichment of chlorinated hydrocarbons, such as 1,1,2-trichloroethane (TCE), trichloroethylene (TCET) and tetrachloroethylene (TECET) from dilute aqueous solutions by PV was investigated [127]. Novel polymers with high selectivity for these solvents were synthesized by radical polymerization, i.e., glassy copolymers composed of (trimethylsilyl) methyl methacrylate (TMSMMA), and rubbery *n*-butyl acrylate (*n*-BA). The effect of the molar ratio of TMSMMA/*n*-BA on the permeation rate of TCE and the separation factor for TCE/H_2O selectivity was examined. The glass transition temperatures of the copolymers decreased with an increase in *n*-BA content, which resulted in high segmental mobility and thus high diffusivity. The copolymer membrane containing about 70 mol % of *n*-BA showed the highest separation factor, in the range of 600-1000, for TCE. The high selectivity of these copolymer membranes for chlorinated hydrocarbons was mainly attributed to high partition coefficients for chlorinated hydrocarbons.

Silicalite-filled poly (siloxane imide) (PSiI) membranes were prepared for the separation of volatile organic compounds (VOCs) from water by PV. PSiI copolymer was synthesized by polycondensation of 3,3',4,4'-benzophenonetetracarboxylic dianhydride (BTDA) with a siloxane-containing diamine, e.g., PDMS, bis(3-aminopropyl) terminated (PSX), added with 3,3-diaminodiphenyl sulfone (DDS). 2,4,6-triamine pyrimidine (TAP) was added into the casting solution to enhance the compatibility between the polymeric matrix and the filler, silicalite. The surface morphology for the membrane with the addition of TAP differed from that without TAP. The latter seems to be consisting of particles in the membrane surface. The solubility selectivity of the PSiI membranes for chloroform ($CHCl_3$)/water solutions was investigated, and there was a highest value for it around 50 wt % of PSX content. The PV performance of the membranes was studied with the separation of chloroform/water mixture. The silicalite-filled membrane with 120 mm thickness exhibited a high total permeation rate of 280 g/ (m^2 h) with separation factor for $CHCl_3$/H_2O selectivity of 52.2 for chloroform of 1.2 wt % of the chloroform/water mixture [128].

HTPB- PUU- PMMA IPN membrane has been developed for the selective removal of chlorinated VOCs such as 1,1,2,2-tetrachloroethane (TCEN), $CHCl_3$, carbon tetrachloride (CCl_4), TCET in water in very low concentration by PV. IPNs of different PMMA content and also different cross-linking density were used. Since the selective permeation and diffusion of the VOCs through the membrane are dependent on their interaction with the membrane material, their absorption and diffusion behaviors through the membrane were also investigated by swelling the membrane in pure VOCs. The absorption and diffusion behaviors were explained with the help of their solubility parameter data and calculated interaction parameter data of the membrane polymers with the VOCs. From the swelling kinetics data, diffusion coefficients of the VOCs through the membrane were calculated. Diffusion coefficients increased with the increase in cross-linking density and PMMA content in the membrane. In PV experiment, concentrations of chlorinated organic compounds in feed were varied from 100ppm to 1,000ppm. All the three IPN membranes showed excellent separation performances of the chlorinated VOCs from water. One IPN containing PMMA of 26% produced TCET of 88.7% in the permeate and resulted in the permeation rate of 0.2 kg/(m^2 h) and a separation factor of 7,842 for 0.1% aqueous feed at 30°C. All three IPN membranes of different compositions showed the separation performances, viz., permeation rate and separation factor in the order of TCEN< CCl_4 < $CHCl_3$ <TCET [129].

Surfactants are widely used for the remediation of groundwater contaminated with chlorinated solvents. The reuse of surfactants is quite important for the economic feasibility of the remediation process. PV with a PDMS membrane was conducted to simultaneously separate TCE and TCEN from waste solution of the non-ionic surfactant Tween 80. Membrane thickness between 200 and 300 μm did not affect the PV performance. The permeation rate of organic did not increase above the flow rate of 100 mLmin^{-1}. The increase in the feed concentration and feed temperature enhanced permeation rate of organic. Organic electivity reduced significantly with increasing feed temperature. This was due to an increase in the thermal motion of polymer chains. The reduction in the permeation rate was small below surfactant of 1.0 wt % (Tween 80) in feed solution. The permeation rate and selectivity for TCE were higher than those of PCE because an extra micellar portion of TCE was higher than that of TCEN. In simultaneous PV of TCE and TCEN, the permeation rate and organic selectivity decreased due to the competitive absorption of TCE and TCEN. However, the reduction percentage in permeation rate and selectivity was less than 10% in TCE/TCEN of 1/1. With operation for 100 h, over 95% of

TCE and 90% of TCEN were simultaneously removed from 0.5 wt % of Tween 80 solution with TCE and TCEN of 1,000 ppm [130].

The application of an integrated PV process to improve the PV performance of acetic acid/water mixtures was investigated. This integrated PV process was based on a plain PDMS membrane with a hydrophobic ionic liquid composed of a heterocyclic cation and $[PF\ 6]^-$ anions. The hydrophobic ionic liquid was introduced as the third phase between the aqueous phase and the plain PDMS membrane for improving mass-transfer of acetic acid from its aqueous matrix to the PDMS membrane. The primary results indicated that the ionic liquid as an extractant prior to PV was favorable for improving the CH_3COOH/H_2O selectivity and the permeation rate of acetic acid compared with those in a plain PDMS membrane. This performance could be attributed to the incorporation of acetic acid molecules and rejection of water molecules by the ionic liquid prior to PV. Extraction of a real effluent containing acetic acid from an antibiotic pharmaceutical plant was carried out using the above integrated PV, and the results imply that this integrated PV process could be scaled up for recovering acetic acids over its water-rich effluents [131].

Hydroxy terminated polybutadiene (HTPB)-based polyurethane urea (PUU) membranes were synthesized and used for pervaporative recovery of *N*-methyl-2-pyrrolidone (NMP) from dilute aqueous solution. The absorption of NMP in cross-linked PUU membranes was found to vary with the *N*-methyl-2-pyrrolidone concentration in the feed and soft segment content of the membrane. A linear absorption isotherm was observed. The effects of membrane soft segment content, *N*-methyl-2-pyrrolidone concentration, operating temperature and membrane thickness on the PV performance of the synthesized PUU membranes were studied. With decrease in soft segment content of the membrane, the permeation rate decreased slightly but the separation factor for NMP/H_2O selectivity increased to some extent. With increasing in concentration of *N*-methyl-2-pyrrolidone in the feed solution, the permeation rate was improved significantly. Both the permeation rate and the separation factor for NMP/H_2O selectivity increased with increasing in operating temperature [134].

HTPB based porous PUU membranes were prepared. The porosity was developed by incorporation of lithium chloride into polymer matrix with subsequent leaching of the same in hot water. The PV performance of synthesized membranes was studied with aqueous 4-nitrophenol solution as feed. PUU membrane was found to permeate 4-nitrophenol selectively with high separation factors for the organic component. Pore size and number of pores on the surface of the membrane were calculated from SEM image of the membranes, and Effects of pore size and porosity on PV permeation were also investigated [132].

Interpenetrating network (IPN) membrane of HTPB-based PUU and PMMA was developed and used in the PV separation of water/*N, N*-dimethyl formamide (DMF) mixtures of different compositions. At a higher temperature, the permeation of diffusing molecules through the membrane became easier, resulting in an increase of permeation rate of DMF. With the increasing in PMMA content in the membrane the permeation rate of DMF and separation factor for DMF/H_2O selectivity increased. This increase was due to the presence of more polar groups of PMMA. The highest permeation rate of *N, N*-dimethyl formamide was 0.231 kg/(m^2 h) for the PUU-PMMA-3 membrane at 60°C for DMF of 80% in the feed [133].

The separation of propyl propionate/water mixtures by PV using poly(ether-*b*-amide) (PEBA) membranes, which is relevant to aroma compound recovery from dilute aqueous solutions, was studied. Under the experimental conditions tested, the permeate concentration was much higher than the solubility limit, and upon phase separation substantially pure propyl propionate could be achieved. The diffusivity of propyl propionate through the membrane from its dilute aqueous solutions was affected by the solution concentration exponentially. The permselectivity of the membrane for propyl propionate/water separation was mainly derived from its solubility selectivity due to the organophilicity of the membrane. The diffusivity of pure propyl propionate in the membrane was about 29 times higher than the diffusivity of pure water [134].

Hydrophobic alumina and titania micro and mesoporous ceramic membranes were prepared by grafting of $C_6F_{13}C_2H_4Si(OEt)_3$ (C6) molecules and subsequently applied in a pervaporation (PV) process to the removal of hazardous organic solvents (MTBE, EtAc and BuOH) from binary aqueous solutions. The transport and separation properties of investigated membranes were discussed. Additionally, it was found that membrane material and grafting time have an important impact of the pervaporative properties. All tested membranes were selective toward organics. Titania membrane was the most efficient and was characterized by the highest value of PSI and the highest value of permeate flux of organic compounds in water–EtAc (J_{EtAc} = 1.1 kg/ (m^2 h)); PSI_{EtAc} = 140 kg/ (m^2 h) and water–MTBE (J_{MTBE} = 1.0 (kg/m^2 h); PSI_{MTBE}=194 kg/ (m^2 h)) systems. This behavior of titania membrane was also discussed based on the Hansen solubility parameters. In presented work the evaluation of MTBE removal degree with titania modified membrane from water binary mixture in the pervaporation process was discussed [135].

Highly hydrophobic titania, alumina and zirconia porous ceramic membranes were prepared by grafting of $C_6F_{13}C_2H_4Si(OEt)_3$ (C6) and $C_{12}F_{25}C_2H_4Si(OEt)_3$ (C12) molecules and subsequently applied in pervaporation (PV) process for removal of volatile organic compounds (methyl tert-butyl-ether/MTBE and ethyl acetate/EtAc) from binary aqueous solutions. Transport of pure water through pristine and hydrophobized membranes were compared. The transport and separation properties in PV process of investigated hydrophobic membranes were discussed. The influence of the membrane materials, pore size of the membranes as well as conditions of the grafting procedure (time and type of molecules) on the pervaporative properties was presented. All investigated membranes were selective toward organic compounds. The highest efficiency was found for the TiO_2 membranes, independently of the utilized system and applied grafting molecules. Better transport and selective properties were observed for membrane modified by C6 molecules than by C12 ones. TiO_2 and ZrO_2 membranes modified by C6 were characterized by high β separation factors 84 (TiO_2) and 56 (ZrO_2). The least effective membrane was alumina one. Additionally, the influence of contact of different solvents (water, EtAc and MTBE) with the hydrophobic layer was discussed. It was found that in the presence of organic compounds in the aqueous solution, the conformation of hydrophobic chains is changing from tangled to straight one [136].

The PV characteristics of the EVA copolymer membranes with different vinyl acetate (VA) content for recovery of ethyl acetate (EA) from aqueous ethyl acetate solutions were investigated. The separation factor for EA/H_2O selectivity decreased with increasing VA content from 26 wt % (EVA26) to 100 wt % (poly vinyl acetate, PVAc) in EVA copolymer membranes. A maximum permeation rate ($J_{EA, max}$) of ethyl acetate of an EVA membrane with a VA content of 38 wt % for 2.5 wt % aqueous EA solution at 30°C was 550 $g(m^2h)^{-1}$ and the separation factor for EA/H_2O selectivity was to 118. These PV characteristics of EVA membranes were explained in terms of the crystallinity of EVA copolymer and the affinities between EA or water and VA segment in EVA copolymer membrane. Experimental results also showed that both the separation factor for EA/H_2O selectivity and J_{EA} increased with increasing feed temperature and feed concentration [137].

Separation of multicomponent liquid systems containing non-volatile compounds (e.g., salts, sugars or proteins) became an important and interesting case of the application of PV process. The influence of NaCl on the performance of PV removal of methyl acetate (MeAc) from water using hydrophobic membrane was evaluated [138]. The selective and transport properties were determined for a hydrophobic PDMS membrane in contact with pure water, binary water/methyl acetate and ternary water/methyl acetate/NaCl systems. The presence of electrolyte in the feed solution caused simultaneously the decrease of the permeation rate of water and the increase of permeation rate of methyl acetate, thus the increase in the membrane selectivity and the overall performance of the separation process were enhanced. Such behavior of the membrane system can be explained by the salting-out effect and described using the Setschenov empirical equation. In contact with binary mixture (1 wt % MeAc), the separation factor for MeAC/H_2O selectivity was 160 and the permeation rate of MeAc $J_{(MeAc)}$ was 202 g/ (m^2 h) whereas in contact with ternary mixture (1 wt % MeAc, 4 mol NaCl/kg) and J_{MeAc} was 430 g/(m^2 h).

PEBA membranes were prepared by casting a solution on a nonsolvent surface. The effects of the solvent ratio (*n*-butanol/2-propanol), temperature, and polymer concentration on the quality of the membranes were studied. The results show that the film quality was enhanced with increasing 2-

propanol ratio in the solvent. This behavior was related to the reduction of the solution surface tension and the interfacial tension between the solution and nonsolvent. Uniform films were made at a temperature range of 70-80°C and a polymer concentration of 4-7 wt %. The qualities of the films improved with increasing 2-propanol ratio in the solvent. With these membranes, the PV of ethyl butyrate (ETB)/water and 2-propanol/water mixtures was studied, for ethyl butylate/water mixtures, with increasing ethyl butylate content, both the permeation rate and separation factor for ETB/H_2O selectivity increased. However, for 2-propanol/water mixtures, with increasing 2-propanol content, the permeation rate increased, but the separation factor for IPA/H_2O selectivity was decreased. Increasing temperature in a limited range resulted in a decreasing separation factor and an increasing permeation rate [139].

Mohammadi et al. [140] prepared poly(ether-block-amide) membranes by casting a solution on a nonsolvent surface. The effects of the solvent ratio (n-butanol/isopropyl alcohol), temperature, and polymer concentration on the quality of the membranes were studied. The results show that the film quality was enhanced with increasing isopropyl alcohol ratio in the solvent. This behavior was related to the reduction of the solution surface tension and the interfacial tension between the solution and nonsolvent. Uniform films were made at a temperature range of 70–80°C and a polymer concentration of 4–7 wt %. The morphology of the membranes was investigated with scanning electron microscopy. The qualities of the films improved with increasing isopropyl alcohol ratio in the solvent. With these membranes, the pervaporation of ethyl butyrate (ETB)/water and isopropyl alcohol/water mixtures was studied, and high separation performance was achieved. For ETB/water mixtures, with increasing ETB content, both the permeation flux and separation factor increased. However, for isopropyl alcohol/water mixtures, with increasing isopropyl alcohol content, the permeation flux increased, but the separation factor was diminished. Increasing temperature in a limited range resulted in a decreasing separation factor and an increasing permeation flux.

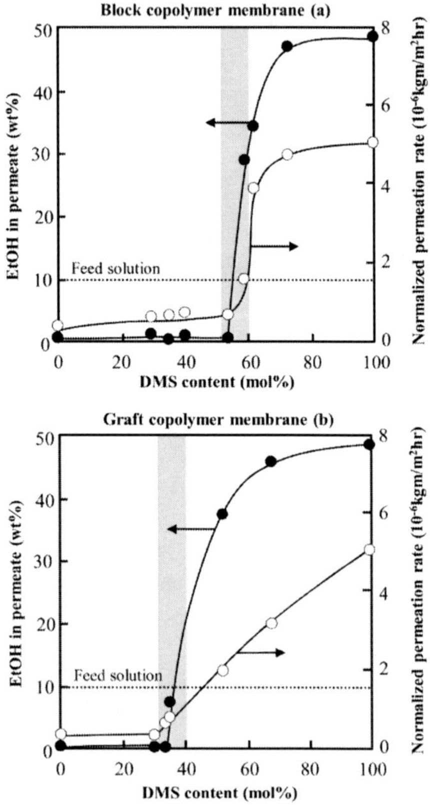

Figure 7 Effects of the DMS content on the ethanol concentration of the permate (●) and on the normalized permeation rate (○) through PMMA-*b*-PDMS (a) and PMMA-*g*-PDMS (b) membranes during pervaporation. The feed solution was an aqueous solution of 10 wt% ethanol (40 °C). The dashed line represents the feed composition.

The removal of volatile organic compounds (VOCs) such as benzene and chloroform from aqueous benzene and chloroform solutions using poly(methyl methacrylate)-poly(dimethyl siloxane) (PMMA-*g*-PDMS), poly(ethyl methacrylate)-PDMS (PEMA-*g*-PDMS), and poly(*n*-butylmethacrylate)-PDMS (PBMA-*g*-PDMS) graft copolymer membranes was investigated by PV as shown in Figures 7 and 8. When aqueous solutions of dilute VOCs were permeated through the PMMA-*g*-PDMS and PEMA-*g*-PDMS membranes, these membranes were Bz/H_2O- and $CHCl_3$/H_2O-selective. The permeation and separation characteristics of the PMMA-*g*-

PDMS and PEMA-*g*-PDMS membranes changed drastically at a DMS content of about 40 and 70 mol %, respectively, as shown in Figure 7. The permeation rate and VOC/water selectivity of the PBMA-*g*-PDMS membranes, however, increased gradually with increasing DMS content, unlike those of PMMA-*g*-PDMS and PEMA-*g*-PDMS membranes. Furthermore, TEM observations revealed that the PMMA-*g*-PDMS and PEMA-*g*-PDMS membranes had microphase-separated structures, consisting of a PDMS phase and a poly (alkyl methacrylate) phase. On the other hand, the PBMA-*g*-PDMS membrane was homogeneous. It was found that the permeability and selectivity of these graft copolymer membranes for treatment of aqueous VOC solutions by PV were significantly related to a PDMS continuous layer in the phase-separated structure [141].

Hydrophobically surface-modified membranes were prepared by adding a fluorine-containing graft copolymer to a microphase-separated membrane consisting of PDMS and PMMA. Contact angle measurements and X-ray photoelectron spectroscopy (XPS) revealed that the addition of a fluorine-containing copolymer produced a hydrophobic surface at the air-side of the membrane due to surface localization of the fluorinated copolymer. It was apparent from TEM that adding a fluorine-containing copolymer of less than 1.2 wt % did not affect the morphology of the microphase-separated membrane. However, adding a fluorine-containing copolymer over 1.2 wt % resulted in a morphological change, from a continuous PDMS phase to a discontinuous PDMS phase. The addition of a small amount of fluorine-containing copolymer to the microphase-separated membranes enhanced both their permeability and Bz/Chx selectivity for a dilute aqueous solution of benzene during PV because of their hydrophobic surfaces and microphase-separated structures. Specifically, the microphase-separated membrane with 1.2 wt % of fluorine-containing copolymer concentrated an aqueous solution from 0.05 to 70 wt % benzene, and, therefore, removed the benzene from water very effectively [141].

Effects of the addition of fluorine-containing graft- or block-copolymer additives composed of PFA and PDMS or PDMS macro-azo-initiator to a microphase-separated membrane consisting of PDMS and PMMA) on the Bz/H_2O selectivity by PV, and the surface characteristics and structures of the surface-modified PMMA/PDMS membranes were investigated Addition of the fluorine-containing copolymer additives, PFA-*g*-PDMS and PFA-*b*-PDMS to a PMMA-*g*-PDMS membrane yielded hydrophobic surfaces at the air-side of surface-modified membranes due to the localization of their additives on the air-side surface of these membranes [142]. The addition of a small amount of the PFA-*g*-PDMS and PFA-*b*-PDMS enhanced both the Bz/H_2O selectivity and the permeability for an aqueous dilute benzene solution during PV because of the formation of their hydrophobic surfaces and the existence of the microphase-separated structures with a continuous PDMS phase in the inner of these membranes. When a larger amount the PFA-*g*-PDMS and PFA-*b*-PDMS were added to a PMMA/PDMS, the latter additive could keep the microphase separated structures with a continuous PDMS phase, but the former did not PMMA-*g*-PDMS and PMMA-*b*-PDMS membranes containing *tert*-butylcalix [4] arene (CA) (CA/PMMA-*g*-PDMS and CA/PMMA-*b*-PDMS) were applied to the removal of benzene from a dilute aqueous solution of benzene by PV. When an aqueous solution of 0.05 wt % benzene was permeated through CA/PMMA-*g*-PDMS and CA/PMMA-*b*-PDMS membranes, these membranes showed high benzene/water selectivity. Both the permeability and Bz/H_2O selectivity of the CA/PMMA-*g*-PDMS and CA/PMMA-*b*-PDMS membranes were enhanced by increasing the CA content, due to the affinity of CA for benzene. The permeability and Bz/H_2O selectivity of CA/PMMA-*b*-PDMS membranes were much greater than those of CA/PMMA-*g*-PDMS membranes. TEM observations revealed that both the CA/PMMA-*g*-PDMS and CA/PMMA-*b*-PDMS membranes had microphase-separated structures consisting of a PMMA phase and a PDMS phase-containing CA. The microphase-separated structure of the latter membranes was much clearer than that of the former and was lamellar. The distribution of CA in the microphase-separated structure of the CA/PMMA-*g*-PDMS and CA/PMMA-*b*-PDMS membranes was analyzed by differential scanning calorimetry (DSC) and CA was distributed in a PDMS continuous layer in micro phase-separated structure [143, 144].

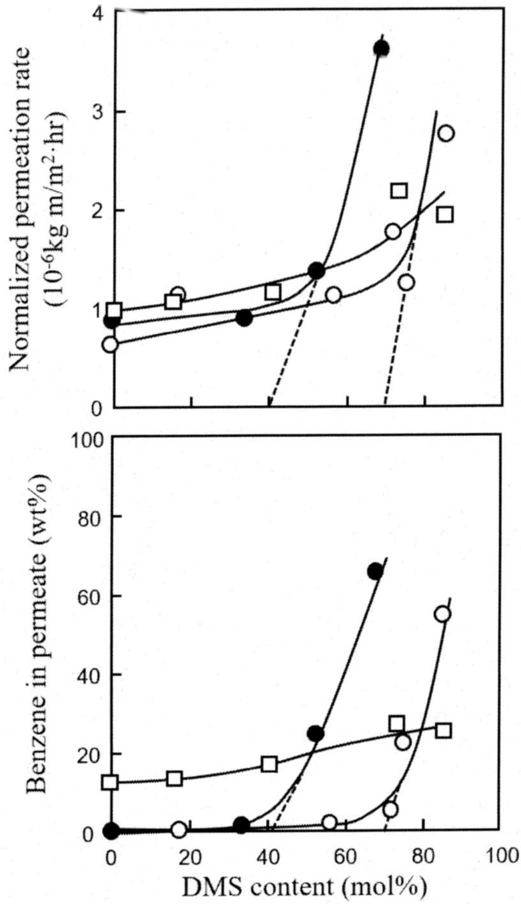

Figure 8. Effect of the DMS content on the benzene concentration in the permeate and normalized permeation rate for an aqueous solution of 0.05 wt % benzene through the PMMA-g-PDMS (●), PEMA-g-PDMS (○), and PBMA-g-PDMS (□) membranes by pervaporation.

In previous works, it has been reported that the PMMA-*b*-PDMS membranes show high VOC/water selectivity. Thus, to improvement of the VOC/water selectivity, graft- and block-copolymer membranes composed of polystyrene (PSt), which is more hydrophobic than PMMA and has higher affinity for VOCs, and PDMS were prepared. On the other hand, recently, ionic liquids are becoming the center in a great number of fields such as inorganic, organic, biological, electrochemical applications because ionic liquids have outstanding properties such as very low vapor pressure, non-volatility, heat-stability, high polarity, good solvent for various organic and inorganic compounds, high conductivity and so on.

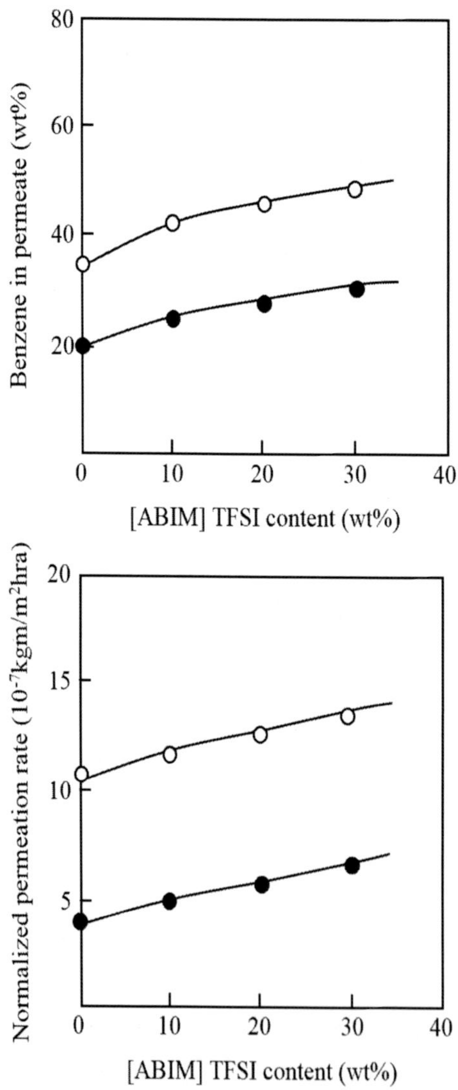

Figure 9. Effect of the [ABIM]TFSI content on the normalized permeation rate (○) and benzene concentration in the permeate (●) for an aqueous solution of 0.05 wt % benzene through the [ABIM]/PMMA-g-PDMS membranes by pervaporation.

Such ionic liquids are playing an increasingly important role in separation science [145]. Then, PSt-*b*-PDMS membranes containing an ionic liquid, 1-allyl-3-butylimidazilium bis (trifluoro methane sulfonyl)imide ([ABIM]TFSI) ([ABIM]TFSI/ PSt-*b*-PDMS) were tested to the removal of VOCs from an aqueous solution of dilute VOC by PV [146]. These [ABIM]TFSI/PSt-*b*-PDMS membranes showed high VOC/water selectivity, and both the permeability and VOC/H_2O selectivity were enhanced by increasing the [ABIM]TFSI content, due to the affinity of [ABIM]TFSI for VOCs as shown in Figure 9. The selective removal of dilute benzene from water by poly (methyl methacrylate)-graft-poly(dimethyl siloxane) membranes containing hydrophobic ionic liquid by pervaporation also was investigated by Uragami et al. [147].

It was found that a continuous PDMS layer in PMMA-*g*-PDMS and PMMA-*b*-PDMS membranes play an important role for the removal of VOCs from water. To construct the membrane matrix from PDMS component mainly, polydimethylsiloxane dimethyl methacrylate macromonomer (PDMSDMMA) was selected as a membrane material. The effects of cross-linkers of the cross-linked PDMS membranes derived from PDMSDMMA and divinyl compounds, on the PV characteristics of the removal of benzene from an aqueous solution of dilute benzene. When an aqueous solution of 0.05 wt % benzene was permeated through the cross-linked PDMSDMMA membranes, they showed high Bz/H_2O selectivity. Both the permeability and Bz/H_2O selectivity of the membranes were enhanced with increasing divinyl compound content as the cross-linker, and were significantly influenced by the kind of divinyl compound. PDMSDMMA membranes cross-linked with divinyl siloxane (DVS) showed very high membrane performance during PV. The best normalized permeation rate, separation factor for Bz/H_2O selectivity, and PV separation index (PSI) [157, 158] which is the product of the permeation rate and the separation factor, and can be used as a measure of the membrane performance during PV, of a PDMSDMMA-DVS membrane were 1.96×10^{-5}mkg/(m^2 h), 98, and 192, respectively [148].

When divinyl perfluoro-*n*-hexane (DVF), which is much more hydrophobic, was employed as a cross-linker of PDMSDMMA, the best normalized permeation rate, separation factor for Bz/H_2O selectivity, and PSI of a PDMSDMMA-DVF membrane were 1.72×10^{-5} mkg/(m^2 h), 4,316, and 7,423, respectively [149].

Effect of the addition of CA to the cross-linked PDMS membranes derived from PDMSDMMA, and various divinyl compounds, on the PV characteristics of the removal of benzene from an aqueous solution of dilute benzene. Both the Bz/H_2O selectivity and permeability of the membranes were enhanced with increasing both divinyl compound content as the cross-linker and additional amount of CA were significantly influenced by the kind of divinyl compound. Addition of CA to PDMSDMMA membranes cross-linked with DVF gave very high effect for the PV performance [150].

In Table 1, the permeation and separation characteristics of various polymer membranes consisting of the PDMS components are compared under the same PV condition: feed solution, an aqueous solution of 0.05 wt % benzene; permeation temperature, 40°C; pressure of permeation side, 1.33Pa. As can be seen in Table 1, both the normalized permeation rate and the Bz/H_2O selectivity of the of the each CA/PDMSDMMA-DVB, CA/PDMSDMMA-DVS and CA/PDMSDMMA-DVF membrane were improved as compared to the each PDMSDMMA-DVB, PDMSDMMA-DVS and PDMSDMMA-DVF membrane. Although, the separation factors of the CA/ PDMSDMMA-DVB, CA/PDMSDMMA-DVS membranes were lower than that of the PFA-*g*-PDMS/PMMA-*g*-PDMS membranes, the PSI of the former membranes was much greater than that of the latter one. In previous studies [148, 149], the introduction of cross-linking structure into the membrane matrix using a suitable cross-linker with a high affinity for permeant was a very effective method to give both a high permeation rate and a high Bz/H_2O selectivity. In Table 1, it found that the addition of CA into the cross-linked PDMSDMMA membranes cross-linked with a suitable cross-linker is significantly effective to obtain a higher permeation and separation characteristics. A CA/PDMSDMMA-DVF membrane with DVF of 90 mol % and CA of 0.4 wt % showed the best membrane performance, i.e., the normalized permeation rate, separation factor for Bz/H_2O selectivity, and PSI were 1.86×10^{-5} mkg/(m^2 h), 5,027, and 9,350, respectively.

Table 1. Performance for Bz/H_2O of various membranes containing PDMS component

Various PDMS Membranes [a]	$\alpha_{sep.}$ Bz/H_2O	$\alpha_{sorp.}$ Bz/H_2O	$\alpha_{diff.}$ Bz/H_2O	NPR [b]	PSI [c]
PMMA	53	422	0.13	0.29	16
PMMA-*g*-PDMS [d]	620	739	0.86	0.13	226
CA/PMMA-*g*-PDMS [e]	1,772	1,267	1.40	0.71	1,240
PFA-*g*-PDMS/ PMMA-*g*-PDMS [f]	4,492			0.64	2,879
PDMSDMMA-DVB [g]	3,171	1,436	2.21	1.46	4,629
PDMSDMMA-DVS [h]	2,886	1,270	2.46	1.96	5,656
PDMSDMMA-DVF [i]	4,316	1,804	2.49	1.72	7,423
CA/PDMSDMMA-DVB [j]	4,021	1,689	2.18	1.75	7,037
CA/PDMSDMMA-DVS [k]	3,866	1,620	2.39	1.97	7,616
CA/PDMSDMMA-DVF [l]	5,027	1,998	2.52	1.86	9,350

[a] PV experimental condition: feed solution, an aqueous solution of 0.05 wt % benzene; permeation temperature, 40 °C; pressure of permeation side, 1×10^{-2} Torr (1.33Pa).
[b] Normalized permeation rate [10^{-5}mkg$(m^2h)^{-1}$].
[c] PV separation index (NPR × $\alpha_{sep.}$ Bz/H_2O).
[d] PDMS content, 74 mol %.
[e] PDMS content, 74 mol %; CA content, 40 mol %.
[f] PDMS content, 74 mol %; PFA-*g*-PDMS content, 1.2 wt %.
[g] DVB content, 80 mol %.
[h] DVS content, 90 mol %.
[i] DVF content, 90 mol %.
[j] DVB content, 80 mol %; CA content, 0.5 wt %.
[k] DVS content, 90 mol %; CA content, 0.5 wt %.
[l] DVF content, 90 mol %; CA content, 0.4 mol %.

The air stripping removal of VOCs such as toluene and phenol from water by microporous poly(propylene) (PP) hollow fibers was studied. The VOC stream passed through the lumen side of the module, while air (stripping gas) flowed across the shell side. Experiments were performed at different liquid flow rates (8-16 cm^3/min), gas flow rates (60-180 cm^3/min), feed VOC concentrations (100-1000 ppm), and temperatures (24-35°C). The removal was more effective when feed VOC level, liquid or gas flow rate increased. The applicability of a mass transfer model that considers diffusion in the liquid layer, membrane, and gas layer under steady state was checked. Unlike phenol with a very small dimensionless Henry's law constant (equilibrium gas concentration divided by liquid concentration) and a relatively low amount of absorption on PP fibers, the measured overall mass transfer coefficients for toluene reasonably agreed with those predicted from the model. The large deviation observed for phenol indicated unsteady state nature, likely due to its small concentration difference between air and liquid phase/fiber matrix. The air stripping removal and separation of binary VOCs including o-xylene and chloroform from water was also examined [151].

The application of pervaporation (PV) to the removal of volatile organic (VOC) from the aqueous solutions has become very interesting in the last few years. It is caused by the increasing level of the compounds such as petrochemical solvents (benzene, toluene, and xylenes) or chlorinated solvents (trichloroethylene or tetrachloroethylene) in the natural environment. The aim of this work is to applythe membranes with carbon black as a filler to PV of toluene from the diluted aqueous solution were applied. PDMS composite membranes, PDMS and PEBA membranes filled with carbon black were used. The membranes were prepared in a laboratory scale especially for this purpose. Both PEBA and PDMS membranes showed excellent properties to separate toluene from the water, however, toluene solubility in the PDMS membrane was better. The removal of this compound from the feed was better when PDMS membrane was used. In some cases total permeation rate as well as toluene permeation rate decreased for the membranes with the filler comparing to those without the filler [152].

The permeation and separation characteristics of VOCs, such as chloroform, benzene, and toluene from water by PV through cross-linked PDMS membranes prepared from PDMSDMMA and divinyl compounds, such as EGDM, DVB, DVS, and DVF were studied [153].

These membranes showed high VOC/H_2O selectivity and permeability. Both VOC/water selectivity and permeability were affected significantly by the divinyl compound. Furthermore, cross-linked PDMSDMMA membranes showed the highest $CHCl_3$/H_2O selectivity. The VOC/H_2O selectivity was mainly governed by the solubility selectivity rather than the diffusion selectivity. However, the difference in the VOC/H_2O selectivity between different types of VOCs depended on differences in the diffusivity of permeants. With increasing downstream pressure, the VOC/H_2O selectivity of all cross-linked PDMSDMMA membranes increased, but the permeability decreased. A PDMSDMMA-DVF membrane exhibited a normalized permeation rate of 1.9 x 10^{-5} kgm/ (m^2 h) It and a separation factor for $CHCl_3/H_2O$ of 4850, yielding a separation index of 9,110.

Atom transfer radical polymerization of HEMA from initiators immobilized on porous supports yields PV membranes with high permeation rate that can be readily modified to control their properties. Derivatization of poly (2-hydroxyethyl methacrylate) (PHEMA) coatings with octyl (C8-PHEMA), hexadecyl (C16-PHEMA), or pentadecafluorooctyl (fluorinated PHEMA) side chains provides films that are sufficiently hydrophobic to allow selective PV of VOCs from water. For all of these derivatized PHEMA membranes, VOC/water selectivity generally increased with decreasing solubility of the VOC in water, and trichloroethylene/water selectivity is about 500. Absorption data suggested that fluorinated PHEMA has the highest free volume and that C16-PHEMA had the lowest, and consistent with this inference, permeation rate decreased in the order of fluorinated PHEMA > C8-PHEMA > C16-PHEMA. The permeation rate was generally an order of magnitude greater than those through high-performance PDMS membranes [154].

Five kinds of PDMS/PVF_2 composite membranes were prepared using asymmetric PVF_2 hollow-fiber membranes as the substrate where a very thin layer of silicone-based coating material was deposited via a developed dip coating method. In the optimal coating procedure, homogenous and stable oligo-PDMS coating layers as thin as 1-2 μm were successfully deposited on the surface of PVF_2 membranes. The developed PDMS/PVF_2 composite membranes were applied for separation of a wide variety of VOC (benzene, chloroform, acetone, ethyl acetate, and toluene). The results showed that the PDMS/PVF_2 hollow-fiber composite membranes that had been developed exhibited very high removal efficiency (> 96%) for all the VOCs examined under favorable operating conditions [155].

The study was aimed at removing organic solvents from aqueous solutions using PV. The tests were carried out on a composite PDMS membrane manufactured by GKSS/Forschungszentrum Geesthacht, Germany and a PDMS membrane filled with carbon black (CB) prepared in a laboratory. There were two runs: the first one tested an aqueous toluene solution as feed, while the other used a mixture of solvents: toluene, acetone and ethyl acetate. The PDMS membrane without carbon black removed all the components of the mixture, displaying a very high efficiency of the process. However, the selectivity and enrichment factors turned out to be relatively low. The PDMS/CB membrane produced slightly lower removal degrees as well as the permeation rate of acetone and ethyl acetate compared to the PDMS membrane, yet their concentration at the permeate side of the membrane was considerably higher. The presence of other solvents did not affect greatly toluene permeation - its final removal seemed to be independent of the composition of the feed used [156].

The syntheses of siloxane-grafted poly(amide-imide) and polyamide were carried out by a macromonomer method, to develop a highly permeable and durable membrane material for PV. A novel compound, 3,5-bis(4-aminophenoxy) benzyloxy propyl-terminated polydimethylsiloxane (BAPB- PDMS), was synthesized as a macromonomer. BAPB-PDMSs with different PDMS segment lengths were prepared by hydrosilylation of 3,5-bis(4-nitrophenoxy) benzyl allyl ether with hydro silyl-terminated PDMS using Pt catalyst, followed by hydrogenation reduction of the terminal dinitro groups. The polycondensations of BAPB-PDMSs with trimellitic anhydride chloride and terephthaloyl chloride yielded the desired siloxane-grafted poly(amide-imide) (PAI-*g*-PDMS) and polyamide (PA-*g*-PDMS) copolymers, respectively. The copolymer membranes were prepared by solvent-casting method, and the gas permeability and PV property of these membranes were evaluated. As the solubility of the copolymers depended on the main chain structure, PA-g-PDMS membrane was insoluble in any solvents after it was dried *in vacuo.* The gas permeability coefficients of these copolymer membranes were increased with increasing of PDMS segment length, and these values of PAI-*g*-PDMS were higher than those of PA-*g*-PDMS containing the same PDMS segment length. From the results of PV of the dilute aqueous solutions of organic solvents, it was found that PA-*g*-PDMS membrane exhibited the excellent organic/water selectivity toward several organic solvents, such as alcohols, acetone, THF, chloroform, dichloromethane and benzene with a high and stable permeation. Furthermore, it was confirmed that the removal of THF from its dilute aqueous solution was efficiently achieved by the PV using PA-*g*-PDMS membrane [157].

3.5. Selective Membranes for Organic Liquid/Organic Liquid Mixtures

Organic/organic selective membranes are effectively used for the purification and separation of industrial products and reuse of organic solvents based on separable recovery of organic mixtures.

3.5.1. Benzene/Cyclohexane Selective Membranes

The separation of benzene and cyclohexane by distillation is a very energy-intensive process, because the boiling points of the components are very similar. PV may be an alternative, more energy-efficient process for the separation of benzene/cyclohexane mixtures. Therefore, many workers have studied the PV properties of polymeric membranes for benzene/cyclohexane separation.

For example, Uragami et al. selected PMMA and PEMA membranes which have a strong affinity for benzene. The PV permeation and separation characteristics for benzene/cyclohexane mixtures in these poly (alkyl methacrylate) membranes cross-linked with ethylene glycol dimethacrylate (EGDM) were studied [158]. The cross-linked poly (alkyl methacrylate) membranes exhibited benzene/cyclohexane (Bz/Chx) selectivity for benzene/cyclohexane mixtures, and the permeation rate increased with increasing benzene concentration in the feed solution. The Bz/Chx selectivity of these membranes was strongly governed by the diffusion selectivity and depended on the difference in molecular size between the benzene molecule and the cyclohexane molecule. With increasing cross-linker content in the membrane, swelling of the membranes was reduced, enhancing the Bz/Chx selectivity. This result was attributed to an increase in the solubility selectivity caused by reduced swelling of the membrane. Cross-linked tertiary copolymer (PMMA-PEMA-EGDM) membranes also showed excellent Bz/Chx selectivity. These results suggested that both the increase in the affinity of the membrane for benzene and the reduction in the swelling of the membrane are very important variables in the separation of benzene/cyclohexane mixtures.

Poly (dimethyl acrylamide-random-methyl methacrylate) (DMAA-*r*-MMA) and poly (dimethyl acrylamide)-graft-poly (methyl methacrylate) (DMAA-*g*-MMA) membranes were also examined for benzene/cyclohexane separation by PV. The benzene selectivity of the DMAA-*r*-MMA membrane changed from diffusion -selectivity-controlled to solubility-selectivity-controlled with increasing DMAA content. In contrast, DMAA-*g*-MMA membranes with higher DMAA content had higher apparent diffusion selectivity than the apparent solubility selectivity. Furthermore, the apparent Bz/Chx solubility selectivity for the DMAA-*r*-MMA membrane and the DMAA-*g*-MMA membrane with a higher DMAA content was remarkably different. These results were attributed to the difference in structure between the copolymers [159].

A side-chain liquid-crystalline polymer (LCP) was synthesized by the addition of mesogenic monomer to poly (methyl siloxane) with a Pt catalyst. When benzene/cyclohexane mixtures were permeated through the LCP membranes by PV at various temperatures, the permeation rate increased with increasing benzene concentration in the feed solution and permeation temperature. Although the LCP membranes exhibited Bz/Chx selectivity, the mechanism responsible for the permeation and separation of the benzene/cyclohexane mixtures was different in the glassy, liquid-crystalline state versus the isotropic state of the LCP membranes. These results suggested that the Bz/Chx selectivity was moderately influenced by the change in LCP membrane structure (i.e., a state transformation). The balance between the orientation of the mesogenic groups and the flexibility of the siloxane chain is very important with respect to permeability and Bz/Chx selectivity [160, 161]. When benzene/cyclohexane, toluene/cyclohexane, and *o*-xylene/cyclohexane mixtures were subjected to PV through a LCP membrane in the liquid-crystalline state, the permeation rate increased with increasing temperature and the LCP membrane exhibited selectivity for the aromatic hydrocarbons. The permeation rate and selectivity of the LCP membrane for each mixture decreased with increasing molecular size of the aromatic hydrocarbon in the binary feed mixture [162]. When benzene/ cyclohexane mixtures were permeated through noematic and sematic side-chain liquid-crystalline polymers (*n*- and *s*-LCP) membranes under various conditions during PV, the *n*- and *s*-LCP membranes exhibited Bz/Chx selectivity. The selectivity of the *n*-LCP membrane changed from solubility-selectivity-controlled to diffusion-selectivity-controlled upon the state-transformation of the membrane, induced by an increase in the permeation temperature. In contrast, the selectivity of the *s*-LCP membrane was governed by diffusion selectivity regardless of the state of this membrane. At low permeation temperatures, the *n*-LCP membrane in the liquid-crystalline state exhibited lower permeability but higher selectivity than the *s*-LCP membrane. However, at high permeation temperatures, the relationship between the permeability and Bz/Chx selectivity of the *n*-LCP and *s*-LCP membranes in the liquid-crystalline state was reversed. These results are a result of differences in the chemical and physical structure of the *n*-LCP and *s*-LCP membranes [163].

Chito and its derivatives have been used as membrane materials for a wide variety of PV applications, including dehydration of ethanol/water mixtures. Uragami et al. synthesized benzoyl chitosan (BzChitos) with different degrees of benzoylation as a durable membrane material for the separation of benzene/cyclohexane mixtures. The characteristics of BzChito membranes, such as contact angle, crystallinity, and the degree of swelling, were significantly influenced by the degree of benzoylation. The BzChito membranes showed high Bz/Chx selectivity for a mixture containing 50 wt % benzene during PV. The difference in Bz/Chx selectivity of the BzChito membranes with different degrees of benzoylation corresponded to differences in the physical structure of the membranes. When a benzene/cyclohexane mixture of 50 wt % benzene was permeated through the BzChito membranes, the permeation rate increased and Bz/Chx selectivity decreased slightly with an increasing degree of benzoylation [164, 165].

Modified cellulose-based membranes for benzene/cyclohexane separation was developed [166, 167]. Various types of benzoyl cellulose (BzCell) were synthesized and the effect of the degree of benzoylation on the PV properties of the membranes was investigated. A BzCell membrane with a benzoylation degree of 2 showed high Bz/Chx selectivity for benzene/cyclohexane mixtures during PV. The permeation rate of the BzCell membrane increased, but the Bz/Chx selectivity decreased with increasing benzene concentration in the feed mixture. This increase in the permeation rate was due to an increase in the swelling of the membrane, whereas the Bz/Chx selectivity decrease was attributed to a decrease in the solubility selectivity. With increasing benzoylation of the BzCells, the permeation rate increased significantly, but the benzene selectivity decreased slightly. These results cannot be explained by the degree of swelling, the density, or the contact angle of the BzCell membranes [168].

Tosylcellulose (TosCell) with different degrees of tosylation were synthesized as membrane materials for the separation of benzene/cyclohexane mixtures. TosCell membranes also showed high Bz/Chx selectivity when tested in PV mode [169]. An increase in the benzene concentration of the feed mixtures increased the permeation rate but decreased the Bz/Chx selectivity of the TosCell membranes. This increase in the permeation rate was attributed to an increase in the degree of swelling of the TosCell membranes due to absorption of the feed mixture. The decrease in Bz/Chx selectivity was mainly caused by a decrease in solubility selectivity. With low benzene concentrations in the mixtures, the permeation rate of a TosCell membrane with a higher degree of tosylation was greater than that with a lower degree of tosylation; however, the reverse was observed for high benzene concentrations. The Bz/Chx selectivity of the former TosCell membrane was higher than that of the latter membrane. Differences in the permeation rate and Bz/Chx selectivity with changes in the benzene concentration of the feed mixture and the degree of tosylation of the TosCell membrane were significantly influenced by the degree of swelling of the TosCell membrane, which is related to the benzene concentration absorbed into the membrane. The mechanism responsible for the separation of benzene/cyclohexane mixtures through the TosCell membranes was analyzed and discussed using the solution-diffusion model.

Methyl methacrylate-methacrylic acid copolymer (P(MMA-*co*-MAA)) membranes ionically cross-linked with Fe^{3+} and Co^{2+} ions P (MMA-*co*-MAA-Fe^{3+} and -Co^{2+}) were prepared, and their permeation and separation characteristics for a benzene/cyclohexane mixture containing 50 wt % benzene by PV were studied. Although the introduction of the metal ions into the MMA-*co*-MAA membrane enhanced both Bz/Chx selectivity and permeability for a benzene/cyclohexane mixture, the PV characteristics of the MMA-*co*-MAA-Fe^{3+} and -Co^{2+} membranes were significantly different. The difference in performance between these membranes was strongly governed by differences in their membrane structures, as indicated by their glass transition temperature, contact angle to methylene iodide, degree of swelling, and mixture composition absorbed into the membrane [170].

Ren et al. investigated the PV properties of a series of cross-linked 4,4'-hexafluoro-isopropylidene dianhydride (6FDA) based copolyimide membranes for the separation of benzene/cyclohexane mixtures [171]. The glassy, highly rigid copolyimides were obtained by polycondensation of 6FDA with various diamines. To obtain high permeability as well as high selectivity, a combination of the diamines 2,3,5,6-tetramethyl-1,4-phenylene diamine (4MPD), 4,4'hexafluoro-isopropylidiene dianiline (6FpDA) and 3,5-diaminobenzoic acid (DABA) as monomers with a cross linkable group were used. Cross-linking is necessary to prevent undesirable swelling effects, which generally occur with non-cross-linked polyimides, especially if high benzene concentrations are reached during PV. The degree of crosslinking was kept constant at 20%, whereas the ratio of the diamine monomers 6FpDA and 4MPD was varied. The PV experiments were performed at 60°C, using benzene/cyclohexane mixtures with benzene concentrations covering the entire concentration range. All of the cross-linked polymers had excellent chemical and thermal stability in the PV experiments. In all cases, conditioning of the membrane samples with pure benzene was a suitable pretreatment to enhance the permeation rate without decreasing the Bz/Chx selectivity significantly. For the most promising membrane material, 6FDA-4MPD/DABA of 4:1 cross-linked with ethylene glycol, the PV experiments were performed with a benzene/cyclohexane feed mixture of 50/50 (w/w) over a temperature range between 60 and 110°C to determine the effect of temperature on the separation characteristics.

The absorption and PV properties of PVC membranes in benzene/ cyclohexane mixtures were studied by Yildirim et al. [172] The effects of the benzene/cyclohexane mixture composition and the temper-ature on the absorption and PV characteristics were determined at 30, 40, and 50°C, respectively, for membranes containing 8 wt % PVC polymers. The total absorption increased with increasing concentrations of benzene. Increasing the concentration of benzene resulted in an increase in the permeation rate and a decrease in the Bz/Chx selectivity. Permeation rates increased, and Bz/Chx selectivity decreased with increasing temperature. The Bz/Chx selectivity was not affected significantly by varying the concentration of polymer in the casting solution, but, as expected for a nonporous membrane, the permeation rate decreased with increasing concentration of the polymer in the casting solution. This was due an increase in membrane thickness.

Polyvinyl acetalized PVA (PVAc) membranes added *O, O*-bis (diethoxy phosphoryl)-*ter*t-butyl calixarene (BEPCA) as a molecular recognition compound were applied to the separation of benzene/cyclohexane mixture. These membranes were high Bz/Chx selective. When evaluated against pure benzene and cyclohexane, PVA/BEPCA showed an ideal selectivity tending towards infinity but a poor permeation rate (5.9 kg μm/(m^{2}h)$^-$ for benzene. The permeation rate of benzene was significantly higher in PVAc and increased from 20 kg μm/(m ^{2}h) for 20% (v/v) benzene/cyclohexane mixture to 65.8 kg μm/(m^2 h) for pure benzene. The swelling index also increased with the increasing concentration of benzene in the feed. PV in PVAc was thus solubility-controlled. The high interaction of PVAc with benzene was interpreted in terms of solubility parameters and free volume in the polymer [173].

The PV characteristics of benzene/cyclohexane mixtures through a fixed carrier membrane consisting of cellulose acetate (CA) as a base polymer and dinitrophenyl (DNP) group as a selective fixed-carrier was investigated. The result of the PV of benzene/cyclohexane mixture using the unmodified cellulose acetate membrane and the CA/DNP membrane showed that the DNP group was effective to increase the Bz/Chx selectivity [174].

Poly (vinyl acetal) PVAc was synthesized by the acid-catalyzed acetalization of poly (vinyl alcohol). Three different morphologically modified variants of the polymer were prepared by the addition of supramolecules; namely α and β cyclodextrins (CD) and butyl calixarene (calix) to the PVAc sol. 20% (w/v) of the solution of the polymers in tetrahydrofuran (THF) was used to cast the membranes. These membranes were characterized using spectroscopic and thermoanalytical techniques, and also tested for their chemical resistance and mechanical strength. The pervaporative behavior of modified membranes vis-à-vis PVAc for the separation of a wide composition of benzene (Bz)/cyclohexane (Chx) mixtures was evaluated. Addition of αCD or calix to PVAc reduced the overall flux while dispersion of βCD favored an enhanced flux especially in the Bz-lean and azeotropic region [175].

The effect of carbon molecular sieve (CMS) incorporated into PVA on PV of benzene/cyclohexane mixtures was studied [176]. The characterization results showed that filling of CMS decreased the hydrogen bond interaction among PVA polymer chains and made chains more flexible and relaxed, which resulted in decreasing of PVA membrane crystallinity, and thus created higher free volume in the membranes. CMS filling effectively increased degree of swelling of membranes in almost whole feed composition range. Due to the more relaxed polymer chain packing and stronger absorption ability, the permeation rate of CMS-filled PVA membranes increased effectively, but excessive filling would decrease the permeation rate. With increase of benzene weight fraction, operating temperature and feed flow rate, the permeation rate of benzene increased, while the separation factor for Bz/Chx selectivity decreased. Compared with unfilled PVA membrane (permeation rate of benzene 21.87 g/ (m^2 h), separation factor 16.7), permeation rate of benzene of PVA-CMS-06 membrane (mass ratio of CMS to PVA was 6%) increased remarkably to 59.25 g/ (m^2 h) and separation factor for Bz/Chx selectivity increased to 23.21.

Novel kind of hybrid membrane was prepared by filling carbon graphite (CG) into PVA and Chito blending mixture, in hopes of improving the separation performance of the membranes by the synergistic effect of blending and filling. The results showed homogenous distribution of graphite particles, considerable alteration of hydrogen bonding interaction, remarkable decrease of crystallinity degree, dramatic enhancement of mechanical properties and significant increase of free volume in CG-PVA/Chito hybrid membranes. Absorption studies indicated that the degree of swelling of CG-PVA/Chito hybrid membranes increased with increasing graphite content and the separation factor for Bz/Chx selectivity was dominated by solubility selectivity rather than diffusivity selectivity. The PV properties of CG-PVA/Chiot hybrid membranes for separation of benzene/cyclohexane mixtures were investigated through varying graphite content and mass ratio of PVA to Chito in hybrid membranes. Higher permeation rate and separation factor of CG-PVA/Chito hybrid membranes were desirably obtained, comparing to PVA and PVA/Chito membranes. A CG06-PVA60/Chito40 membrane exhibited the highest separation factor of 59.8 with a permeation rate of 124.2g$(m^2h)^{-1}$ at 50°C, 1 kPa [177].

β-CD-filled cross-linked PVA membranes were cross-linked with GA (β-CD/PVA/GA). The β-CD/PVA/GA membranes showed strong Bz/Chx selectivity. The permeation rates of β-CD/PVA/GA membranes increased when the β-CD content was 0-8 wt %, but the permeation rate decreased slightly more than β-CD content of 8 wt %. The separation factor for Bz/Chx selectivity increased when β-CD content was in the range 0-10 wt % and decreased slightly in the range of β-CD content of 10-20 wt %. Compared with the β-CD-free PVA/GA membrane, the separation factor for Bz/Chx selectivity of the β-CD/PVA/GA membrane considerably increased from 16.7 to 27.0, and the permeation rate of benzene increased from 23.1 to 30.9 g/(m^2 h) for benzene/cyclohexane (50/50 w/w) mixtures at 50°C. The analysis based on the solution-diffusion model and β-CD inclusion phenomena (complex stabilities) revealed that β-CD in PVA membranes played an important role as a carrier to partition benzene selectively from benzene/cyclohexane mixtures [178].

Novel hybrid membranes were prepared from PVA and carbon nanotube (CNT) of β-CD [179]. Both pure PVA and PVA/CNT hybrid membranes were uniform and these hybrid membranes exhibited significant improvement in Young's modulus and thermal stability as compared to pure PVA and β-CD/PVA membranes. These membranes were applied to PV separation of benzene/cyclohexane mixtures, and showed excellent PV properties. The permeation rate of benzene was 61.0 g/ (m^2 h) and separation factor for Bz/Chx selectivity was 41.2.

An aliphatic type of hyperbranched-polyester (HBPE) that was synthesized from 2,2-dihydroxymethyl propanil acid (DMPA) and 1,1,1-trihydroxymethyl propane (TMP) was end-capped with acrylic group, and the photo-polymerizable hyperbranched-polyester (AHBPE) was obtained [180]. These AHBPEs were employed as the macromolecular cross-linking agents to enhance the PV performance of ethyl cellulose (EC) membrane for the separation of benzene/cyclohexane mixtures. The effects of the content of AHBPE and benzophenone (BP) as well as the radiating time on the separation factor and the total permeation rate were investigated in detail. The results show that, the performance of EC membranes incorporating AHBPEs of higher generations exhibited much higher permeation rate, in a comparison to the membranes cross-linked with ethylene glycol methacrylate. A membrane from AHBPE of 40 wt % and BP of 10 wt % showed the total permeation rate of 42.5 kg μm/ (m^2 h) and the separation factor for Bz/Chx selectivity of 6.82.

Novel hybrid membranes were prepared by incorporating H-(3-zeolite or Rh-loaded H-β-zeolite (Rb/H-β-zeolite) into polyvinyl chloride (PVC) for the pervaporative separation of benzene and cyclohexane [181]. The membranes were characterized by scanning electron microscopy (SEM) and X-ray diffraction (XRD) to assess their morphology. The characteristics of these membranes for separating benzene/cyclohexane mixtures were investigated by varying zeolite content, feed composition and operating temperature. The results showed a typical trade-off curve; fluxes increased and selectivity decreased with the increase of benzene concentration in feed composition. The pervaporation performance for membranes filled with Rh/H-β-zeolite was much better than that for pure PVC membrane and membranes filled with H-β-zeolite. It might be caused by the special interaction between benzene and Rh/H-β-zeolite found by IGC previously. And the membrane containing 7 wt % of Rh/H-β-zeolite showed the highest separation selectivity of benzene at all feed compositions. From the temperature-dependent permeating values, the Arrhenius activation parameters were estimated. The resulting lower: apparent activation energy values for permeation flux obtained for the Rh/H-β-zeolite-filled membranes compared to the H-β-zeolite-filled membranes were attributed to the special interaction between Rh/H-β-zeolite and benzene.

Functionalized multiwalled carbon nanotubes (MWNTs) were incorporated into a chitosan membrane for separation of benzene/cyclohexane mixtures by pervaporation. The pristine MWNTs were treated by mixed acid (H_2SO_4/HNO_3=3/1) and then functionalized with isonicotinic acid. Ag^+ was grafted to the pyridine ring on the MWNTs by a complexation reaction. Membranes made of chitosan and Ag+/carbon nanotubes were prepared with polysulfone membranes as support. The surface structure of modified MWNTs was examined by FT-IR, TEM, EDS and XPS to assess its morphology. The swelling and pervaporation characteristics of the membranes for separation of benzene/cyclohexane mixtures were investigated. The swelling degree of the membranes in benzene was more than 2.5 times that of the pristine chitosan membrane. The swelling degree was found to increase with the content of the MWNTs-Ag^+ in the membrane and the benzene content in the feed. Pervaporation results also showed that the MWNTs-AgVchitosan hybrid membrane separation performance is better than MWNTs/chitosan hybrid membrane and pristine chitosan membrane. The permeation flux of the membrane increased, the selectivity increased firstly and then decreased with the increase in the content of MWNTs-Ag^+. The selectivity reached 7.89 and the permeate flux was 357.96 g/m^2 h (the mass ratio of benzene was 50%, the temperature was 20°C) when the percentage by weight of the MWNTs-Ag^+ to chitosan was 1.5%. The Arrhenius activation parameter of 1.5% MWNTs-Ag^+/chitosan (CS) hybrid membrane is 4.28 kJ/mol while 1.5% MWNTs/CS membrane is 10.6 kJ/mol in 50% benzene concentration in feed [182].

3.5.2. Organic/Organic Selective Membranes

The preferential absorption and PV permselectivity of a poly (vinyl chloride) (PVC) membrane for various binary liquid mixtures were evaluated [183]. Methanol/1-propanol, benzene/*n*-hexane, and ethanol/water mixtures were selected as model mixtures in this study. For the methanol/*n*-propanol mixture, methanol preferentially absorbed in the PVC membrane, resulting in selective methanol permeation. For the benzene/*n*-hexane mixture, benzene exhibited higher absorption uptake and permeated preferentially. In the ethanol/water mixture, ethanol preferentially absorbed in the PVC membrane, but water was the preferentially permeating component. This result suggested that the overall selectivity in the membrane was determined by a high water/ethanol diffusion selectivity. The absorption data were analyzed according to Mulder's model [184] derived from Flory-Huggins thermodynamics. The PV selectivity in these systems was analyzed using a solubility and diffusion selectivity model.

Composite membranes of polystyrene (PS) and a hydrophilic polymer were prepared via the concentrated emulsion polymerization method [185]. In the concentrated emulsion precursor, styrene containing styrene-butadiene-styrene block copolymers (SBS) constituted the dispersed phase, and a solution of a hydrophilic monomer in water was the continuous phase. The polymerized systems were transformed into membranes by hot pressing at 150°C. The mechanical properties of the membranes were affected by the amount of SBS in the dispersed phase and by the nature of the hydrophilic monomer of the continuous phase. SBS improved the mechanical properties of the membrane significantly. The membranes were then subjected to absorption and PV of benzene/ethanol mixtures. They exhibited preferential absorption of benzene over the entire benzene concentration range. The swelling ratio increased with increasing benzene concentration, and the solubility selectivity decreased. These membranes had permeation rates as high as 1,040 g/ (m^2 h), and separation factors as high as 25 for benzene/ ethanol mixtures

Zhou et al. developed composite membranes based on a selective polypyrrole layer for the separation of ethanol and cyclohexane [186].

Polypyrrole films were deposited on stainless steel meshes by anodic electro-polymerization of pyrrole dissolved in acetonitrile. Electrochemical and morphological studies on the growth of the polypyrrole films, both oxidized with PF_6^- as the counter-ion, and neutral polypyrrole membranes were conducted. The performance of these membranes for hx separation of ethanol/cyclohexane mixture by PV was investigated. The results indicated preferential permeation of ethanol, and clearly demonstrated the feasibility of exploiting conducting polymers in the PV process.

The PV characteristics for acetone/chlorinated hydrocarbon mixtures using blend membranes of natural rubber (NR) with epoxidized NR (ENR). The permeation rate and selectivity of these membranes were determined both as a function of the blend composition and of the feed mixture composition. The membrane performance could be optimized by adjusting the blend composition. NR/ENR 70/30 and NR/ENR 30/70 compositions showed a decrease in permeation rate and chlorinated hydrocarbon/acetone selectivity, whereas the 50/50 composition showed an increased permeation rate and increased hydrocarbon/acetone selectivity. Chlorinated hydrocarbons permeated preferentially through all of the tested membranes. The feed mixture composition also strongly influenced the PV characteristics of the blend membranes. The chlorinated hydrocarbon/acetone selectivity was found to depend on the molecular size of the penetrants [187].

The permeation behavior of methanol and methyl acetate (MAC) in the PV experiments are compared with those in VP experiments using a PVA-based composite membrane. The selected membrane was found to be MeOH/MAc selective. Results showed higher permeation rate but a similar separation factor for MeOH/MAc selectivity in PV than in VP. For PV operation, the resulting separation factor for MeOH/MAc selectivity at 60°C showed a monotonous decrease (6.4-4.1) with increasing methanol concentration in the feed mixture (2.3-34 wt %), whereas the total permeation rate increased from 0.97 to 7.9 kg/ (m^2 h). Based on the solution-diffusion theory, a mathematical model that described satisfactorily the permeation rates of methanol and methyl acetate in both the PV and VP processes was applied. The permeation rates of both permeants could be explained by the solution-diffusion model with variable diffusion coeffic-ients dependent on methanol concentration in the membrane. Both PV and VP processes can be described with the same model but using different fitting parameters [188].

Copolymers of methyl acrylate and acrylic acid were synthesized to fabricate membranes ionically crosslinked using aluminum acetylacetonate for the separation of toluene/*i*-octane mixtures by pervaporation at high temperatures [189]. The formation of the ionic crosslinking via bare aluminum cations was characterized by UV–VIS spectroscopy and solubility tests. Reproducibility and the reliability of the methodology for membrane formation and crosslinking were confirmed. The effects of acrylic acid content, crosslinking conditions, pervaporation temperature, and feed composition on the normalized flux and the selectivity for toluene/*i*-octane mixtures were determined. A typical crosslinked membrane showed a normalized flux of 26 kg μm/ (m^2 h) and a selectivity of 13 for a 50/50 wt % feed mixture at 100°C. The pervaporation properties including solubility selectivity and diffusivity selectivity are discussed in terms of swelling behavior. The performance of the current membranes w benchmarked against other membrane materials reported in the literature.

The transport properties of hybrid membranes based on polyvinyl-chloride (PVC) was studied for the separation of Toluene–*n*-Heptane mixtures by pervaporation (PV) [190]. Aromatic-alkane mixtures are difficult to fractionate by conventional processes and PV has already been shown as a promising technology in this case. PVC was chosen as starting material in this study because it gathers two key advantages: firstly, it is a technical polymer which is readily available, conversely to special polymers often developed for membrane application; secondly, PVC is a polar polymer endowed with a good affinity for aromatic solvents. Thus mixed matrix PVC membranes were studied to determine if it was possible to obtain interesting separation properties from this simple and cheap polymer. We report the results obtained with several types of clay used as fillers to get hybrid PVC membranes. Hence, PVC based mixed matrix membranes were prepared with Maghnite H, Maghnite H^+, Wyoming, Kaolin and Nanocor clay particles. It was found that the PVC transport properties could be drastically modified both by the amount and by the type of clay incorporated. According to the fillers incorporated in the polymer matrix, it was shown that the transport properties could be easily tuned either as barrier materials or as Toluene selective membranes with strongly enhanced flux compared to initial PVC membranes.

Pervaporation could be a cost-competitive process for separating aromatic/aliphatic hydrocarbon mixtures in the chemical industry. Commercial poly(ether-*block*-amide) (PEBA) and ceramic tubular substrates were used to prepare pervaporation membrane for separating aromatic/aliphatic mixtures through a facile thermal crosslinking method [191]. The absorption and diffusion properties of toluene and *n*-heptane in PEBA were determined by swelling experiment and inverse gas chromatography technique. FTIR and XRD were used to investigate the hydrogen bonding and crystallinity of PEBA molecule through thermal crosslinking. The morphology and structure of PEBA/ceramic composite membrane were characterized by SEM and EDX. The separation performance of the composite membrane can be adjusted by controlling PEBA concentration, thermal crosslinking temperature and time. Ceramic tubular substrate could suppress the excessive swelling and thereby enhance the stability of the membrane. Furthermore, different types of fillers were incorporated into PEBA matrix to improve the separation performance. The results demonstrate that PEBA/graphite hybrid membrane has the highest separation factor for separating 50 wt % toluene/*n*-heptane mixtures. This study is expected to extend the membrane material and simplify the membrane preparation procedures in aromatic/aliphatic pervaporation fields.

As a new type of filler for nanohybrid membranes, metal–organic frameworks (MOFs) have attracted intense interest in recent years. In this work, MOF-based $Cu_3(BTC)_2$/PVA (BTC=benzene-1,3,5-tricarboxylate, PVA=poly (vinyl alcohol)) nanohybrid membranes were fabricated on a ceramic tubular substrate by using a pressure-driven assembly method [192]. The morphology and structure of the resulting membranes were characterized by scanning electron microscope, energy dispersive X-ray spectrometer, and powder X-ray diffraction. The $Cu_3(BTC)_2$/PVA membranes were then used in separating 50 wt % toluene/*n*-heptane mixtures through pervaporation. The effects of PVA concentration, $Cu_3(BTC)_2$

loading, feed composition, and operating temperature on membrane performances were explored. The results indicate that, compared with pristine PVA membrane, the separation factor and permeate flux of optimized $Cu_3(BTC)_2$/PVA membranes are improved from 8.9 and 14 g/(m^2 h) to 17.9 and 133 g/(m^2 h), respectively. A speculation of the transport process of permeating components in the selective layer of $Cu_3(BTC)_2$/PVA membrane was proposed. Enhanced affinity between toluene and the membrane through incorporating $Cu_3(BTC)_2$ particles plays a key role in improving separation performances.

Co(II)-formate ($Co(HCOO)_2$) has been demonstrated feasible to be used as stable aromatic hydrocarbon transport carrier for facilitating aromatic hydrocarbon transport membranes. Metal-organic framework Co $(HCOO)_2$ materials with different particle sizes were prepared through a solvothermal synthetic method, and then were doped into poly(ether-block-amide) (PEBA). The resulting hybrids were finally deposited on the outer surface of tubular ceramic substrate with a dynamic pressure-driven assembly method. The morphologies and structures of the resulting $Co(HCOO)_2$/ PEBA membranes were characterized by scanning electron microscope (SEM), energy dispersive X-ray spectrometer (EDX), powder X-ray diffraction (PXRD), thermogravimetric analysis (TGA), positron annihilation lifetime spectroscopy (PALS), and nanoindentation. These membranes were then used in separating aromatic/aliphatic hydrocarbon mixtures through pervaporation. When the feed solution is 10 wt % toluene/*n*-heptane mixtures, the permeate flux of the membrane is 771 g/ (m^2 h) with a separation factor of 5.1. Meanwhile, due to the immobilization effect of the Co $(HCOO)_2$ for Co (II) ions, the Co $(HCOO)_2$/PEBA membrane showed stable separation performances [193].

Pervaporation is a potentially attractive process for the separation of dimethyl carbonate (DMC)/methanol mixture. Li et al. [194] have studied the cross-linking density, swelling degree and pervaporation performance of the DMC-selective hydrophobic nano-silica/polydimethylsiloxane membrane by varying the membrane preparation parameters including cross-linking agent concentration, nano-silica loading, solvent concentration, operating time and temperature. The structural morphology, thermal stability and mechanical properties of the membrane were also investigated. The results indicate that the thermal and mechanical stability of membrane can be enhanced by the well distributed silica incorporation, and the membrane containing 15 wt % hydrophobic nano-silica has the highest separation factor of 3.97 with a permeation flux of 702 g/(m^2 h) for 30 wt % DMC at the feed temperature of 40°C. Additionally, the absorption and diffusion selectivity of the membrane were measured and discussed via the solution–diffusion mechanism. The very low calculated diffusion selectivity of the membrane indicates that diffusion is the limiting factor in pervaporation performance.

New composite membrane with a poly(γ-benzyl-L-glutamate) (PBG) separation layer on a micro-porous poly(amide-imide) support was prepared and tested as a pervaporation membrane for selective separation of toluene/*n*-heptane mixtures [195]. Composite membrane was formed by coating a solution of PBG in chloroform on the top surface of micro-porous PAI-SO_2 support. PBG layer fulfils indispensable requirement for the preferential absorption of the most penetrating component (toluene) of a separated mixture.

The structure of the membrane was studied in detail. X-ray diffraction, electron microscopy and IR spectroscopy measurements were performed. Crystalline regions formed by backbones of macromolecules are low permeable barriers to penetrants, but also limiters for swelling of high permeable polymer phase. The knowledge of the material structure in dependence of conditions of its preparation enabled to adapt the process of the membrane formation to reach the efficiency of separating aromatic/ aliphatic compounds.

The separation ability was tested by pervaporation measurements for different toluene concentrations in binary mixtures toluene/*n*-heptane. Diffusion coefficients, fluxes and separation factors were evaluated. It was shown that conformation of α-helix PBG molecules in the coating layer of the composite membrane does not change in the pervaporation. The membrane retains its integrity and can be used in subsequent cycles of a process.

The obtained results during this study could be used for the design of highly efficient PBG/PAI-SO_2 composite membranes for separation of mixtures aromatic and aliphatic hydrocarbons by pervaporation.

The separation of aromatic/aliphatic mixtures is significant in chemical industry. Pervaporation has attracted increasing attention due to its low energy consumption and environmentally friendly process. However, the formation of membranes mostly use toxic organic solvents, whereas in this study water was used instead to prepare membrane in separating aromatic/aliphatic mixtures. W3000 were dispersed in water to form micelles, which were then deposited onto the surface of the tubular porous ceramic substrate through the negative pressure-driven assembly method. The emulsion and composite membranes were characterized by TEM, SEM and FTIR. The crosslinking density of W3000 was also tested using NMR XLD analyzing system. The membranes were used for separating aromatic/aliphatic mixtures through pervaporation. The results showed that compared with the membrane prepared by DMF, the pervaporation separation index (PSI) of the composite membrane formed using water as solvent was 3.3-fold increased for separating toluene/*n*-heptane mixtures. The as-prepared composite membranes showed a better comprehensive pervaporation performance [196]. This facile strategy may have a potential in the application of pervaporation for separating aromatic/aliphatic mixtures in industry.

Natural rubber (NR) and polystyrene-*co*-butadiene) rubber (SBR) were cross-linked chemically with sulfur and accelerator by efficient vulcanization. These were then compounded for physical cross-linking with high abrasion carbon black filler with three different doses, i.e., 5, 10 and 20phr to obtain three filled membranes from each of these two rubbers (NR-5, NR-10 and NR-20 and SBR-5, SBR-10 and SBR-20). These six filled rubber membranes were used for PV separation of toluene (Tol)/methanol mixtures up to 11 wt % of toluene in feed. With increase in filler doses from membrane-5 to membrane-20, the selectivity and mechanical properties of the membranes increased. All these membranes showed reasonably good range of permeation rate (20.8 g/(m^2 h) for SBR-5 to 10.7 g/(m^2 h) for NR-20) and separation factor for Tol/MeOH selectivity (286.4 for NR-5 to 183.7

for SBR-5) for 0.55 wt % of toluene in feed. Among these membranes NR-20 and SBR-20 with highest cross-linking density showed maximum separation factor for Tol/MeOH along with good permeation rate. For comparable cross-linking density NR membranes showed better separation factor than SBR membranes [197].

Ionic liquids are used as a bulk liquid membrane for separation of toluene from *n*-heptane (Hep). Aromatic hydrocarbon, toluene, is successfully transported through the ionic liquid membrane based on 1-methyl-3-octyl imidazolium chloride. Using silver ion as a carrier in 1-methyl-3-octyl imidazolium chloride membranes, batch wise extraction experiments are carried out. The separation performances, represented by the permeation rate and separation factor for Tol/Hep selectivity, were analyzed systematically by varying the operating parameters: the contact time, concentration of Ag^+, stirring effect, initial feed phase concentration and temperature [198].

PV separation of ternary and quaternary mixtures of ethyl acetate, water, ethanol and acetic acid, which are present in the hydrolysis of ethyl acetate encountered in pharmaceutical industries was investigated. Water concentrations in the feed mixtures are in the range of 90-98 wt %, while the concentrations of ethyl acetate, ethanol and acetic acid are much lower. PDMS was used as membrane to separate organic compounds from aqueous streams. PV experiments showed that PDMS membrane was much more selective to ethyl acetate than other organic components. Increase in the ethyl acetate concentration in feed mixture resulted in higher total permeation rates but lower selectivity of ethyl acetate [199].

The blend membranes with different PAA content were used to separate dimethyl carbonate (DMC)/methanol azeotrope in PV. The blend membrane containing 70 wt % PAA had the highest separation factor for DMC/MeOH selectivity of 13 and a permeation rate of 577 g/ (m^2 h) at 60°C for separating an azeotrope [200].

Copoly (diethyl vinyl benzyl phosphonate/hydroxyethyl methacrylate) [copoly (VBP-HEMA)] and copoly (diethyl vinyl benzyl phosphonate/vinyl benzyl chloride) [copoly (VBP-VBC)] were prepared and their membranes cross-linked by a reaction of EGDE and ED with, respectively, hydroxyl and methylene chloride groups in their side chains. PV and absorption of benzene/*n*-hexane, benzene/cyclohexane, and toluene/*n*-octane mixtures were investigated [201]. The membranes were in a rubbery state and preferentially permeable to aromatics. They had higher specific permeation rate and a lower PV separation factor compared with the cross-linked membranes of methacrylate copolymers with analogous pendant groups. They displayed better PV performance for toluene/*n*-octane. Absorption isotherms of benzene/*n*-hexane mixtures were represented by the Flory-Rehner model.

Different types of membranes based on polyelectrolytes were developed for the separation of aromatic/aliphatic hydrocarbon mixtures by PV. For this purpose composite membranes with different polyelectrolytes, soluble either in water (Sulfoethylcellulose) or in ethanol (custom-made copolymers of MMA and methacrylic acid [3-sulfo-propyl ester] potassium salt) were prepared by an interfacial reaction with oppositely charged ionic surfactants. The latter copolymer could also be, cross-linked with Co^{2+} ions. Both membrane types showed a clear enrichment of aromatic components (toluene or benzene) during the PV process (20 wt % in the feed and about 55 wt % in the permeate). The permeation rates were in the range of 1 kg/(m^2h) at 80°C. The separation of a benzene/cyclohexane mixture was also successful at 50°C. In all cases, permeation rates in benzene/cyclohexane mixtures were higher than those Also, in multi-component feed mixtures, benzene exhibited the better permeate rates [202].

AgCl/PMMA organic/inorganic hybrid membrane has been fabricated by reverse microemulsion polymerization using MMA as oil phase [203]. The membrane structure is analyzed by scanning electron microscope (EM), which showed that core-shell organic-inorganic hybrid materials were formed with AgCl as the core and PMMA as the shell. AgCl particulate under 2 μm are uniformly dispersed in the PMMA. The swelling-absorption behavior of the cyclohexane and cyclohexene in AgCl/PMMA hybrid membrane were measured, the absorption capacity of cyclohexene in the hybrid membrane was enlarged comparing with pure PMMA membrane, but the change of the absorption behavior of cyclohexane was less. The equilibrium swelling-absorption amount ratio of cyclohexene to cyclohexane in membranes was up to 130.4, which was larger than 54.7 in the pure PMMA membranes.

The possibility of using ionic liquids in bulk (non-supported) and supported liquid membranes for the selective transport of organic molecules was demonstrated. A systematic selective transport study, in which 1,4-dioxane, 1-propanol, 1-butanol, cyclohexanol, cyclohexanone, morpholine, and methyl morpholine was served as a model seven-component mixture of representative organic compounds, and in which four ion liquids based on the 1-*n*-alkyl-3-methylimidazolium cation (*n*-butyl, *n*-octyl, and *n*-decyl) are used together with the anions PF_6- or BF4-, immobilized in five different supporting membranes, confirms that the combination of the selected ion liquid with the supporting membranes was crucial to achieve good selectivity for a specific solute. The use of the 1-*n*-butyl-3-methylimidazolium hexafluorophosphate, immobilized in a PVF_2 membrane, allows an extremely highly selective transport of secondary amines over tertiary amines (up to a 55:1 ratio). The selective transport of a given solute through the ion liquid/membrane system resulted from the high partitioning of the solute to the liquid membrane phase [204].

As part of a research project focusing on the development of a sustainable biocatalytic process for production of chiral secondary alcohols, the pervaporative separation of ethanol from ethanol/ethyl acetate/1-methoxy-2-propanol/1-methoxy-2-propyl acetate-mixtures through a commercial PVA-based membrane was investigated. PV of the non-diluted multicomponent mixture was shown to be strongly influenced by interactions between the permeants and the membrane. Investigation of these interactions contributed to the understanding of the mass transport mechanism of this mixture. Overall, high permeation rates were obtained, but small differences between the fastest permeating species were found. The fastest permeating species was ethanol, ethyl acetate or 1-methoxy-2-propanol depending on the feed composition [205].

Whether as a membrane solution can be utilized for hydrocarbon separation by using a supported liquid membrane was investigated. Aromatic hydrocarbons, benzene, toluene and *p*-xylene were successfully transported through the membrane based on the ionic liquids. Although the permeation rates through the membrane based on the ionic liquids were less than those of water, the selectivity of aromatic hydrocarbons was greatly improved. The maximum selectivity to heptane was obtained using benzene in the aromatic permeation and 1-*n*-butyl-3-methylimidazolium hexafluorophosphate in the liquid membrane phase [206].

PDMS composite membrane was used to separate the binary and multicomponent alkane/thiophene mixtures by PV. The total permeation rates for different alkane/thiophene mixtures decreased with increase of carbon number in the alkanes. In PV results of ternary systems the increase of content of lighter alkane in feed resulted in a larger total permeation rate, but a smaller selectivity to thiophene. A quaternary system, the mixture of n-heptane, n-octane, n-nonane and thiophene, was employed to simulate the desulfurization process of gasoline. With the membrane having a PDNTS layer of 11 mμm, the total permeation rate was measured to be about 1.65 kg/ (m^2 h), with the corresponding separation factor for thiophene of 3.9 at 30°C [207].

The blend membranes from a polymer blend of PAA and PVA were evaluated for the separation of methanol from methyl *tert*-butyl ether (MTBE) by PV. Methanol permeated preferentially through all blend membranes tested, and the MeOH/MTBE selectivity increased with increasing PVA content in the blends. However, a decrease in the permeation rate was observed with increasing PVA content. Upon increasing the feed temperature, the permeation rate increased and the MeOH/MTBE selectivity remained constant. In addition, the influence of cross-linking on the MeOH/MTBE selectivity was investigated. The PV permeation rate decreased with increasing cross-linking density; however, this was coupled with an increase MEOH/MTBE selectivity. This was due to a more rapid decrease in the partial permeation rate of MTBE compared to that of methanol [208].

Thin-film composite membranes were prepared Thin-film composite membranes by coating a thin film of polystyrene sulfonate (PSS) onto the surface of a microporous alumina (Al_2O_3) support [209]. The PSS/Al_2O_3 composite membrane was applied to the separation of methyl *tert*-butyl ether/methanol mixtures in PV mode and exhibited excellent performance. The PSS/Al_2O_3 composite membranes showed high MeOH/MTBE. The methanol concentration in the permeate was always greater than 99.5 wt %

for all membranes tested and for all feed solution compositions studied. Membranes of the Mg^{2+}-counter ion form (PSS-Mg) showed higher separation factors for MeOH/MTBE selectivity than membranes having Na^+ as the counter ion (PSS-Na). Extremely high separation factors (25,000 to 35,000) were observed for the PSS-Mg/Al_2O_3 composite membrane containing 27.5 mol % sulfonate.

The PV properties of a poly (urethane imide) block-copolymer membrane for separation of ethyl *tert*-butyl ether/ethanol mixtures was investigated by Jonquieres et al. over the entire composition range [210]. The PV selectivity for ethanol was higher than the simple solubility selectivity of ethanol over ethyl *tert*-butyl ether. The features of absorption were examined in relation to the corresponding PV properties, and were analyzed in terms of the activity of each penetrant to consider the non-ideality of the system. Absorption isotherms could be expressed in terms of a single linear relationship for ethyl *tert*-butyl ether but required two linear relationships in the case of ethanol. The discontinuity in the ethanol isotherm corresponded to the composition of a critical feed mixture, and was ascribed to the complete mono-solvation of the polymer's most basic site, i.e., urethane groups, by the protic penetrant.

The composite membranes with PVA as separating layer material and PAN or cellulose acetate (CA) as supporting layer material were prepared for separating methyl *tert*-butyl ether/methanol mixture by PV. The results showed that PV performance of the composite membrane with PVA membrane as separating layer was superior to that with CA membrane as separating layer, and the PV performance of PVA/CA composite membrane with CA membrane as supporting layer was better. The parameters to prepare the composite membrane remarkably affected PV performance of the composite membrane. The permeate rate of both composite membranes of PVA/PAN and PVA/CA was over 400 $g(m^2h)^{-1}$, and the concentration of methanol in the permeate reached over 99.9 wt % for separating methyl tert-butyl ether/methanol mixture [211].

Polyelectrolyte-surfactant complex (PELSC) membranes were prepared by the interaction of the anionic groups of sodium cellulose sulfate (Na-CS) or sulfoethyl cellulose (SEC) with different types of cationic surfactants. The membrane properties were investigated using PV experiments with different organic feed mixtures, containing methanol as a polar component. Methanol was mixed with cyclohexane, methyl *tert*-butyl ether, and dimethyl carbonate. The results of the separation experiments employing the methanol-containing mixtures showed that the membranes reveal both very high permselectivity and permeability even at moderate temperatures [212].

Absorption and PV experiments were carried out with cellulose triacetate membranes for the separation of methyl *tert*-butyl ether and methanol mixtures. In the PV experiments, the total and methanol permeation rates increased with the increasing methanol concentration in the feed while the permeation rates of methyl *tert*-butyl ether first increased and then decreased. The total permeation rates are significantly enhanced with increasing temperature. This temperature dependence was more pronounced at low methanol concentration, but the extent of increase of the total permeate rate was relatively constant when methanol concentrations are greater than 10%. The MeOH/MTBE selectivity decreased with increasing of methanol concentration and became more or less constant at high methanol concentrations [213].

Absorption and PV experiments were carried out with PVA/PAA cross-linked membranes for the separation of an azeotropic methyl *tert*-butyl ether/methanol mixture. With increasing PAA content in the membranes, solubilities and permeation rates decreased and MeOH/MTBE selectivity increased. Total absorption and permeation rates increased with increasing concentration of methanol. Increasing the concentration of methanol resulted in decreasing MeOH/MTBE selectivity. Because of polarity, methanol permeated selectively through the membranes. Absorption results showed the same tendency with PV results [214].

PV separation process was used for separation of methanol/methyl *tert*-butyl ether mixtures encountered during manufacturing of methyl *tert*-butyl ether. Three different copolymers of acrylamide with increasing amount of 2-hydroxyethyl methacrylate (HEMA), i.e., PAMHEMA-1, -2 and -3 were synthesized and the cross-linked (gelled) copolymer membranes made from these sol copolymers were used for pervaporative separation of methanol from its mixtures with methyl *tert*-butyl ether over the concentration range of 0-10 wt % methanol in feed. These hydrophilic gel copolymer membranes were found to be highly methanol selective in both absorption and diffusion through the membranes. The permeation rate of methanol was found to decrease with increasing degree of cross-linking from PAMHEMA-1 to -3 copolymer membranes. Among the three membranes the PAMHEMA-3 membrane showed the highest MeOH/MTBE selectivity and permeation rate and they were 511.7 and 9.9 g/(m^2 h), respectively for 0.53 wt % of methanol in feed, while maximum permeation rate with high selectivity was shown by PAMHEMA-1 membrane. Their permeation rate and MeOH/MTBE selectivity were 18.49 g/ (m^2 h) and 243, respectively for 0.53 wt % of methanol in feed. Methyl *tert*-butyl ether was also found to have a negative coupling effect on methanol permeation [215].

The separation of ethanol/ethyl-tertiobutylether mixtures by PV was studied with new membranes prepared from *N*-vinyl-pyrrolidinone (NVP) and *N*-[3-(trimethylamoniopropyl)] methacrylamidemethylsulfate) (TMA). The PV results showed that highly ethanol selective membranes could be obtained from PVP blends and from pyrrolidinone-based cross-linked copolymers. Whatever the exact NVP/TMA composition used, the membranes strongly favored the PV of ethanol. The ethanol selectivity was higher for the lower PVP/TMA ratio. On the one hand, these results were ascribed to the high pyrrolidinone residues content, which is responsible of the enhanced ethanol absorption affinity. The observed permeation selectivity was in agreement with the swelling data also recorded with the different polymers, showing higher affinity for ethanol with PVP-enriched materials compared with TMA ones. This was a direct consequence of the Lewis base feature of pyrrolidinone sites towards ethanol molecules. On the other hand, the TMA residues improved the overall stability and selectivity of the membranes depended upon crosslinking reactions, which were induced by thermal treatment. A close comparison made between polymer blend and copolymer PV results helped to clarify the TMA role of the membrane transport properties [216].

3.6. Selective Membranes for Isomers

Butanol, propanol and xylene isomer selective membranes are very important for the separation of industrially chemical products and energy saving for the separation of those mixtures.

Mixed matrix membranes (MMMs) comprising polyamide–imide (PAI) and α-, β- or γ-cyclodextrin (CD) have been investigated experimentally and computationally for isomeric *n*-butanol/tert-butanol (*n*-BuOH/*t*-BuOH) separation via pervaporation [217]. Consistent with molecular simulation, experimental results show that the CD inclusion ability and butanol discrimination ability are dependent on both CD cavity size and butanol molecular size. The PAI membrane incorporated with α-CD has the smallest cavity and has the highest discrimination ability for the *n*-BuOH/*t*-BuOH pair but with a low butanol flux. The mixed matrix membrane embedded with γ-CD has the lowest selectivity and the highest flux. The PAI/β-CD membrane has a comparable selectivity and flux and exhibits preferential absorption and diffusion selectivity toward n-BuOH. A maximum separation factor of 1.53 with a corresponding flux of 4.4 g/ (m^2 h) are obtained at an optimal β-CD loading of 15 wt %. Further increments in the CD content eventually lead to a decrease in separation performance because of CD agglomeration and severe phase separation. To better understand the influence of CD on the separation performance of mixed matrix membranes, SEM, FTIR and XRD have been employed for membrane characterizations. The effect of n-butanol/t-butanol ratio in the feed composition has also been studied. It is found that both flux and separation factor decrease with increasing n-butanol content in the feed. The decline is attributed to the change in total vapor pressure at the upstream and the mutual drag effect of isomeric butanol molecules.

The separation of xylene isomers was carried out by pervaporation through a polymer membrane having a dinitrophenyl group. The polymer membrane was effective for selective permeation of m-xylene from the mixture of xylene isomers. This selectivity can be attributed to the difference in the charge-transfer complexing ability between xylene isomers and the dinitrophenyl group [218].

PVA membranes containing β-CD (PVA/CD membrane) were prepared and the permeation and separation characteristics for propanol isomers through the PVA/CD membranes were investigated by PV and EV [219]. EV was more effective for the separation of propanol isomers through the PVA/CD membrane than PV. The PVA/CD membrane preferentially permeated 1-propanol rather than 2-propanol from their mixtures. In particular, a mixture of 10 wt % 1-propanol concentration was concentrated to about 45 wt % through the PVA/CD membrane. The permeation mechanism of propanol isomers through the PVA/CD membrane was discussed based on the solution-diffusion model.

The characteristics of permeation and separation for xylene isomers through PVA membranes containing β-CD (PVA/CD membranes) were investigated by PV and EV. EV was more effective for the separation of xylene isomers through the PVA/CD membrane than PV. The increase in CD content gave an increase in *p*-xylene/*o*-xylene selectivity through the PVA/CD membrane by EV. This was attributed to the stronger affinity of CD for *p*- xylene compared with *o*- xylene. Especially, the PVA/CD membrane at a CD content of 40 wt % showed a higher separation factor for *p*- xylene/*o*-xylene selectivity than has ever been reported. When the *p*-xylene concentration in the feed was lower, the *p*- xylene/*o*- xylene selectivity was improved. The mechanism of permeation and separation for xylene isomers was discussed from the standpoint of solution-diffusion model [220].

PAA membranes containing α-, β-, or γ- CD were prepared and used for the separation of *o*- /*p*-xylene mixture by PV. The native PAA membrane was almost impermeable for the xylene isomers, and the incorporation of CDs in the PAA membranes resulted in membranes having molecular recognition function, which selectively facilitated the transport of the xylene isomers. For all types of CDs, the facilitated transport occurred at CD concentrations higher than the threshold concentration. As the CD concentration increased, the permeation rates increased, while the *o*-xylene/*p*-xylene selectivity were almost constant. The selectivity for *o*-xylene/*p*-xylene selectivity of the membranes was strongly influenced by the types of CDs incorporated in the membranes [221].

Polymer membranes containing α-CD were prepared by the casting method using cross-linking reaction with hexamethylenediisocyanate. The film synthesis conducted with and without dibutyltin dilaurate as catalyst, resulted in two series of materials in which α-CD host entities were chemically linked to PVA physically entrapped, respectively. The obtained membranes were successfully applied to the separation of *o*- / *p*- and *o*- / *m*-xylene isomer mixtures by pertraction from water. *p*- and *m*-xylenes were found to be the faster permeants compared to the *o*- isomer. The separation factor for *p*-xylene /*o*-xylene selectivity varied from 7.75 to 0.35 depending on the membrane α- CD content and the feed concentration. The permeation rate and separation selectivity data were discussed in terms of molecular recognition by α-CD and of coupling transport effect [222].

The PV characteristics of xylene isomer mixtures through a fixed carrier membrane consisting of CA as a base polymer and dinitrophenyl (DNP) group as a selective fixed-carrier were studied. In the PV of xylene isomer mixtures, the DNP group selectively facilitated the transport of xylene isomers through the membrane. The order of the preferentially permeating component was *p*-xylene > *m*-xylene > *o*-xylene [223]. The selective extraction of the geometrical isomers of xylene using a hydrogel consisting in a PVA grafted with α-CD and β-CD as the complexing moieties was described. The membrane contactors were prepared by the casting method following cross-linking reaction with hexamethylene diisocyanate (HMDI). The transfer of xylenes across the CD-containing membranes was facilitated compared to PVA. The better discrimination was observed for membranes based on α-CD, the more efficient being that containing ca 21 wt % CD. The order of permeation rate *p*-xylene > *m*-xylene > *o*-xylene followed the affinity order inferred from the stability constants α-CD. The permeate composition was independent upon the feed composition [224].

Fractionation of *o*- and *p*-xylene isomeric mixtures was performed using PV with PU membranes containing ZSM zeolite. The xylene vapor absorption isotherms exhibited a Henry's law relationship in this PU- zeolite blend. In binary solutions the individual xylene uptake was also proportional to the solvent composition. Although incorporating zeolite into the PU-zeolite membrane rendered a decrease in xylene solubility as compared with that absorbed in the PU film without zeolite addition, the increase of diffusion coefficient and diffusivity selectivity increases enhanced the separation efficiency using the PU-zeolite blend. Increasing the operating temperatures enhanced the xylene permeation rate of xylene. The permeation rates of xylene and selectivity increased with increasing zeolite content [225].

Palygorskite-polyacrylamide (PGS/PAM) hybrid materials were synthesized via intercalation polymerization initiated by redox initiator consisting of modified PGS (reducer) and ceric salt Ce^{4+} (oxidant), and used as PV membranes. The swelling behavior of hybrid membranes was investigated in single xylene isomer (*p*-xylene, *m*- xylene and *o*-xylene), binary xylene isomer mixtures (*p*-/*o*-xylene (the mixtures of p- and o-xylene), o- /*m*- xylene (the mixtures of *o*- and *m*- xylene) and *p*- /*m*- xylene (the mixtures of *p*- and *m*-xylene)) and ternary isomer mixture (*p*- /*m*- /*o*-xylene (the mixtures of *p*-, *m*- and *o*-xylene)). A maximum value of degree of swelling at equilibrium (DS equilibrium) in single xylene isomer at 30°C exhibited for the hybrid membrane with a PGS of 1.92 wt %. Negative deviation and both negative and positive deviations of the DS equilibrium based on the addition rule existed in binary or ternary xylene isomer mixtures respectively for hybrid membranes with different PGS content at 30°C. Also, a maximum value of separation factor of the hybrid membrane revealed for each pair of binary xylene isomer mixtures when the PGS content was 1.92 wt % in the hybrid membrane. A reversion of the preferential selectivity and a high permeation activation energy of the hybrid membrane occurred at the concentration region with high *p*-xylene content in *p*- /*m*-xylene binary xylene isomer mixtures. The swelling behavior and PV performance of PGS-PAM hybrid membranes were discussed in terms of the entrapping channel in the PGS for the potential introduction of xylene isomers, the interaction between xylene isomers in feed and the solution-diffusion mechanism in the PV process [226].

Mixed matrix membranes (MMMs) comprising polyamide–imide (PAI) and α-, β- or γ-cyclodextrin (CD) have been investigated experimentally and computationally for isomeric *n*-butanol/*tert*-butanol (*n*-BuOH/*t*-BuOH) separation via pervaporation [227]. Consistent with molecular simulation, experimental results show that the CD inclusion ability and butanol discrimination ability are dependent on both CD cavity size and butanol molecular size. The PAI membrane incorporated with α-CD has the smallest cavity and has the highest discrimination ability for the n-BuOH/t-BuOH pair but with a low butanol flux. The mixed matrix membrane embedded with γ-CD has the lowest selectivity and the highest flux. The PAI/β-CD membrane has a comparable selectivity and flux and exhibits preferential absorption and diffusion selectivity toward n-BuOH. A maximum separation factor of 1.53 with a corresponding flux of 4.4 g/ (m^2 h) are obtained at an optimal β-CD loading of 15 wt %. Further increments in the CD content eventually lead to a decrease in separation performance because of CD agglomeration and severe phase separation. To better understand the influence of CD on the separation performance of mixed matrix membranes, SEM, FTIR and XRD have been employed for membrane characterizations. The effect of n-butanol/t-butanol ratio in the feed composition has also been studied. It is found that both flux and separation factor decrease with increasing n-butanol content in the feed. The decline is attributed to the change in total vapor pressure at the upstream and the mutual drag effect of isomeric butanol molecules.

Polyvinyl alcohol (PVA) membrane filled β-cyclodextrin (β-CD) was prepared by casting an aqueous solution of PVA and β-CD oligomer. The membrane was crosslinked with glutaraldehyde for one hour. The weight content of β-CD in the membrane was 33%. The membrane was used for separation of *p*-xylene/*m*-xylene mixture by pervaporation. Based on the experiments of absorption equilibrium, the solubility and the diffusion coefficient of the permeates in the membrane was obtained. Compared with PVA membrane, the solubilities of pure *p*-xylene and *m*-xylene in PVA membrane filled β-CD increased from 0.92, 0.78 to 10.4, 2.6 g (xylene)/100 g (dried membrane), respectively, and the solubility selectivity S_p/S_m increased from 1.18 to 4.0. Also, the diffusion coefficients of *p*-xylene and *m*-xylene decreased from 8.45×10^{-12}, 8.23×10^{-12} to 6.83×10^{-12}, 7.23×10^{-12} m^2/s, respectively, and diffusion selectivity decreased from 1.03 to 0.94. These effects of β-CD can be interpreted in terms of the inclusion strength in the cavity. The pervaporation performance of the PVA membrane filled β-CD was investigated. A separation factor of 2.96 and a permeation rate of 95 g/ (m^2 h) through the β-CD filled PVA membrane for a 10 wt % feed *p*-xylene concentration was obtained at 25°C. The results indicated that PVA membrane filled β-CD effectively improved the pervaporation performance, especially on the separation factor [228].

Silicalite-1 zeolite was modified with 3-aminopropyltriethoxysilane (APTES) and then loaded into poly (acrylic acid) sodium (PAAS) to prepare PAAS/silicalite-1 hybrid pervaporation membranes for the separation of xylene isomer mixtures [229]. Characterization by Fourier transform infrared (FT-IR) and nuclear magnetic resonance (NMR) spectroscopies confirmed that chemical modification on the zeolite surface had taken place. Based on absorption equilibrium experiments, the diffusion coefficients of permeates in the membrane were obtained, and the order of the diffusion coefficients of xylene isomers was found to be $D_o < D_m < D_p$. An investigation of the effects of the original versus modified zeolite loading on the pervaporation performance was performed. With the original zeolite, the permeation flux of the binary xylene mixtures showed little change, whereas the selectivity of the hybrid membrane exhibited some enhancement. After modification of the zeolite surface, the selectivity of the hybrid membrane was clearly improved, the permeation flux of the membrane decreased slightly, and the maximum separation factors of the *p*-/*o*-xylene ($\alpha_{p/o}$) and *p*-/*m*-xylene ($\alpha_{p/m}$) mixtures were determined to be 2.62 and 2.68, respectively. The experimental results revealed that the modification promoted the compatibility of the PAAS and silicalite-1 zeolite interface.

Molecularly imprinted polymeric membranes (MIPMs) were prepared from cellulose and 1,2- dihydroxy benzene as a print molecule with imprinting ratio of 0.5 and 1.0 [230]. Those membranes were applied to the separation of xylene isomers by pervaporation. The molecularly imprinted membranes selectively incorporated *o*-xylene from *o*-/*m*- and *o*-/*p*- mixtures at a low *o*-xylene concentration region, the absorption selectivity toward *o*-xylene was determined to be 7.15 and 4.24, respectively, implying that 1,2-dihydroxybenzene worked well as the print molecule for *o*-xylene recognition; the pervaporation was carried out at 40°C and at the down-stream pressure of 0.5 6.6 kPa and showed permselectivity toward *o*-xylene at low *o*-xylene concentrations while toward *m*- and *p*-xylene at high *o*-xylene concentrations. The molecularly imprinted membranes gave higher flux than the control non-imprinted membrane. The results obtained in the present study revealed that molecular imprinting is one of promising methods to prepare separation membranes for pervaporation.

3.7. Facilitation of Chemical Reactions by Pervaporation Membrane

Selective membranes for facilitation of chemical reactions are useful for reactors coupled with membrane separation technology such as PV, EV, TDEV. These systems can help to enhance the conversion of reactants for thermodynamically or kinetically limited reactions via selective removal of one or more product species from reaction mixtures.

3.7.1. Facilitation of Esterification

Author greatly recommend to lean "3.6 Pervaporation Membrane Reactor" written by Professor B. Van der Bruggen in "Comprehensive Membrane Science and Engineering" (Second Edition) [231]. A parametric study was carried out to provide a fundamental understanding of the reactor behavior. A batch reactor integrated with a pervaporation unit was selected as the model system. It was shown by simulation that conversions exceeding equilibrium limits can be achieved by using pervaporation to remove water from the reaction mixtures, and that complete conversion of one reactant is obtainable when the other is in excess. The membrane reactor tolerates the presence of water, which can be either in the reaction medium or as impurity of the reacting reagent. There are upper and lower limits in the performance of reactor facilitation by pervaporation. Membrane permeability, membrane area and the volume of the reaction mixtures to be treated are important operating parameters influencing the reactor behavior. Operating temperature influences reactor performance through its influences on reaction rate and membrane permeability.

The pervaporation-facilitated esterification was investigated by Feng and Huang [232] and a parametric study was carried out to provide a fundamental understanding of the reactor behavior. A batch reactor integrated with a pervaporation unit was selected as the model system. It was shown by simulation that conversions exceeding equilibrium limits can be achieved by using pervaporation to remove water from the reaction mixtures, and that complete conversion of one reactant is obtainable when the other is in excess. The membrane reactor tolerates the presence of water, which can be either in the reaction medium or as impurity of the reacting reagent. There are upper and lower limits in the performance of reactor facilitation by pervaporation. Membrane permeability, membrane area and the volume of the reaction mixtures to be treated are important operating parameters influencing the reactor behavior. Operating temperature influences reactor performance through its influences on reaction rate and membrane permeability.

The core part of this work deals with the review of the most promising studies about pervaporation membrane reactors (PVMRs) for esterification. After presenting the Fischer esterification reaction, the conventional reaction separation technologies are described. The most relevant pervaporative technologies defined in the last 25 years are then shown considering catalyst used, configurations, membranes materials, performances, and overall conversions obtained. Moreover the sustainability and the flexibility of the PVMR is remarked. As for the membranes, polyvinyl alcohol membranes showed to be the most assessed membranes, being cheap and very flexible, whereas zeolite-type membranes gave high water selectivity and permeance, leading to a high yield of the ester produced (these membranes mainly suffer from low reproducibility). In this chapter, some focus is also given to multilayered membranes, bifunctional multilayered membranes (which are completed with a catalytic layer that increases the reagents conversion and, together with the water permeation driving force, water flux), and biocatalytic (enzyme catalyst) reactors. Finally, a concise overview of ester-selective PVMR configuration research activities is also provided [233].

The esterification reaction of propionic acid with isobutyl alcohol to produce isobutyl propionate and water was studied by in the present study, the esterification reaction of propionic acid with isobutyl alcohol to produce isobutyl propionate and water was studied by Rathod et al. [234]. The performance of esterification reaction was compared by using the batch process and the pervaporation assisted hybrid process which performs the reaction and separation simultaneously. A polyvinyl alcohol–polyether-sulphone (PVA–PES) hydrophilic polymeric membrane was used in the study to separate water and also to shift the equilibrium. The influence of process parameters such as catalyst loading, molar ratio of acid to alcohol, reaction temperature and ratio of membrane area to initial reaction volume (S/V) was studied. The results showed that the pervaporation assisted esterification process gave more conversion than the batch process of esterification. The membrane showed high selectivity to the removal of water in the propionic acid, isobutyl alcohol, isobutyl propionate and water mixture. Moreover, the conversion of propionic acid was enhanced by enhancing the catalyst amount, molar ratio of acid to alcohol, reaction temperature and S/V ratio. The performance of esterification reaction was compared by using the batch process and the pervaporation assisted hybrid process which performs the reaction and separation simultaneously. A polyvinyl alcohol–polyethersulphone (PVA–PES) hydrophilic polymeric membrane was used in the study to separate water and also to shift the equilibrium. The influence of process parameters such as catalyst loading, molar ratio of acid to alcohol, reaction temperature and ratio of membrane area to initial reaction volume (S/V) was studied. The results showed that the pervaporation assisted esterification process gave more conversion than the batch process of esterification. The membrane showed high selectivity to the removal of water in the propionic acid, isobutyl alcohol, isobutyl propionate and water mixture. Moreover, the conversion of propionic acid was enhanced by enhancing the catalyst amount, molar ratio of acid to alcohol, reaction temperature and S/V ratio.

Poly (vinyl alcohol) (PVA) membranes crosslinked with sulfosuccinic acid (SSA) were used for the esterification of acetic acid by isoamylic alcohol [235]. In order to study the effects of the crosslinking degree and, simultaneously, the amount of sulfonic groups, different membranes were prepared with SSA: PVA ratios in the range of 5–40 mol %. Aiming at to eliminate the dependence between the amount of acid sites and the crosslinking degree, were also prepared PVA membranes in which the –SO_3H groups were introduced by anchoring 5-sulfosalisilic acid (SA) on the PVA chains.

The conversion of isoamylic alcohol increases when the amount of sulfosuccinic acid used in the polymer crosslinking is increased from 5 to 20%. However, when crosslinking degree increases from 20 to 40%, the conversion of isoamylic alcohol increases only slightly, probably due to the increase of molecules mobility restrictions, in the PVA matrix. In the case of the PVA membranes where the –SO_3H groups were introduced by esterifying 5-sulfosalisilic acid on the PVA –OH groups, it was observed that membrane activity increases with the polymer crosslinking. It was also observed an increase of the activity of these membranes with the amount of –SO_3H groups in the polymeric matrix.

Catalytic stability of PVA membranes, prepared with SSA and SA, was evaluated by performing consecutive batch runs with the same membrane being observed, after the third run, a trend to stabilization of catalytic activity.

Esterification of lactic acid and succinic acid with ethanol to generate ethyl lactate ($C_5H_{10}O_3$) and diethyl succinate, respectively, was studied in well-mixed reactors with solid catalysts (Amberlyst XN-1010 and Nation NR50) and two PV membranes (GFIF-1005 and T1-b). Experiments were carried out by a closed-loop system of a “batch” catalytic reactor and a PV unit employing GFIF-1005. The kinetics of PV is studied to obtain a working correlation for the permeation rate of water in terms of temperature and water concentration on the feed side of the pervaporator. The efficacy of PV-aided esterification was illustrated by attainment of near total utilization of the stoichiometrically limiting reactant within a reasonable time. Protocols for recovery of ethyl lactate and diethyl succinate from PV retentate were discussed and simultaneous esterification of lactic and succinic acids, which is an attractive and novel concept, is proposed [236].

Glycerol ketals are often considered as important bio-based diesel additives or perfume substances which could be synthesized from the acetalization of glycerol and cyclohexanone, however, acetalization is a typical example of reactions controlled by thermodynamic equilibrium. In this study, a catalytically active membrane was prepared by immersion phase inversion and was used for the conversion enhancement of present reaction by continuous removal of by-product water in a pervaporation membrane reactor. SEM images showed that a highly porous "sponge-like" catalytic layer immobilized with catalyst Zr $(SO_4)_2 \cdot 4H_2O$ was uniformly coated on a polyvinyl alcohol/polyethersulfone pervaporation membrane. Synthesis performances in batch reactor, inert membrane reactor and catalytically active membrane reactor were compared, showing that no equilibrium limitation for glycerol conversion was observed in both the inert membrane reactor and catalytically active membrane reactor. The effect of various operational parameters on the synthesis performance in the catalytically active membrane reactor were investigated, showing that higher temperature and A/V ratio favored better enhancement in glycerol conversion since an acceleration of water removal rate was induced. Optimal catalyst loading and initial mole ratio was also determined for better synthesis performance. Under optimized conditions, the glycerol conversion reached 93% and a conversion enhancement of approximately 52% was achieved when compared to equilibrium conversion at 75C [237].

The combination of the chemical reaction step with a pervaporation process can increase the conversion of reversible reactions such as esterification by removing selectively the water formed from the reacting mixture. The esterification of acetic acid with isopropanol was carried out in a reactor combined with a pervaporation unit. The conversions achieved are distinctly higher than the equilibrium conversion. Kinetic and pervaporation parameters obtained in a previous study were used to describe the behavior of the hybrid process. The influence of different operating parameters such as reaction and pervaporation temperature, ratio of membrane area to initial reaction volume, initial molar reactant ratio and amount of catalyst on the process performance has been analyzed in this work [238].

Copolymers of methyl acrylate and acrylic acid were synthesized to fabricate membranes ionically crosslinked using aluminum acetylacetonate for the separation of toluene/*i*-octane mixtures by pervaporation at high temperatures [239]. The formation of the ionic crosslinking via bare aluminum cations was characterized by UV–VIS spectroscopy and solubility tests. Reproducibility and the reliability of the methodology for membrane formation and crosslinking were confirmed. The effects of acrylic acid content, crosslinking conditions, pervaporation temperature, and feed composition on the normalized flux and the selectivity for toluene/*i*-octane mixtures were determined. A typical crosslinked membrane showed a normalized flux of 26 kg μm/ (m^2 h) and a selectivity of 13 for a 50/50 wt % feed mixture at 100°C. The pervaporation properties including solubility selectivity and diffusivity selectivity are discussed in terms of swelling behavior. The performance of the current membranes w benchmarked against other membrane materials reported in the literature.

The transport properties of hybrid membranes based on polyvinylchloride (PVC) was studied for the separation of toluene–*n*-heptane mixtures by pervaporation (PV) [240]. Aromatic-alkane mixtures are difficult to fractionate by conventional processes and PV has already been shown as a promising technology in this case. PVC was chosen as starting material in this study because it gathers two key advantages: firstly, it is a technical polymer which is readily available, conversely to specialty polymers often developed for membrane application; secondly, PVC is a polar polymer endowed with a good affinity for aromatic solvents. Thus mixed matrix PVC membranes were studied to determine if it was possible to obtain interesting separation properties from this simple and cheap polymer. We report the results obtained with several types of clay used as fillers to get hybrid PVC membranes. Hence, PVC based mixed matrix membranes were prepared with Maghnite H, Maghnite H^+, Wyoming, Kaolin and Nanocor clay particles. It was found that the PVC transport properties could be drastically modified both by the amount and by the type of clay incorporated. According to the fillers incorporated in the polymer matrix, it was shown that the transport properties could be easily tuned either as barrier materials or as Toluene selective membranes with strongly enhanced flux compared to initial PVC membranes.

Pervaporation is a good option to enhance conversion in reversible esterification reactions, generating water as a byproduct. Sonawane et al. [241] used polyvinyl alcohol-polyether sulfone (PVA-PES) composite hydrophilic membrane for pervaporation-assisted esterification of lactic acid with *n*-butanol. Pervaporation reactor is an esterification reactor combined with pervaporation membrane module. Experimental work of esterification of lactic acid with *n*-butanol coupled with pervaporation was carried out. Effect of various parameters, such as, initial mole ratio of n-butanol over lactic acid, the ratio of the effective membrane area over the volume of reacting mixture, process temperature and catalyst concentration on the performance of pervaporation reactor were discussed. The following optimum conditions were obtained: temperature = 90°C, catalyst concentration = 0.422 kmol/m^3, initial molar reactant ratio (n-butanol/lactic acid) = 1.4, ratio of effective membrane area to the reaction volume = 15.19 m^2/m^3. The presented data can be extended for study and design of pervaporation reactor for similar kind of reactions.

Pervaporation could be a cost-competitive process for separating aromatic/aliphatic hydrocarbon mixtures in the chemical industry. Commercial poly(ether-*block*-amide) (PEBA) and ceramic tubular substrates were used to prepare pervaporation membrane for separating aromatic/aliphatic mixtures through a facile thermal crosslinking method [242]. The absorption and diffusion properties of toluene and *n*-heptane in PEBA were determined by swelling experiment and inverse gas chromatography technique. FTIR and XRD were used to investigate the hydrogen bonding and crystallinity of PEBA molecule through thermal crosslinking. The morphology and structure of PEBA/ceramic composite membrane were characterized by SEM and EDX. The separation performance of the composite membrane can be adjusted by controlling PEBA concentration, thermal crosslinking temperature and time. Ceramic tubular substrate could suppress the excessive swelling and thereby enhance the stability of the membrane. Furthermore, different types of fillers were incorporated into PEBA matrix to improve the separation performance. The results demonstrate that

PEBA/graphite hybrid membrane has the highest separation factor for separating 50 wt % toluene/*n*-heptane mixtures. This study is expected to extend the membrane material and simplify the membrane preparation procedures in aromatic/aliphatic pervaporation fields.

As a new type of filler for nanohybrid membranes, metal–organic frameworks (MOFs) have attracted intense interest in recent years. In this work, MOF-based $Cu_3(BTC)_2$/PVA (BTC=benzene-1,3,5-tricarboxylate, PVA=poly (vinyl alcohol)) nanohybrid membranes were fabricated on a ceramic tubular substrate by using a pressure-driven assembly method [243]. The morphology and structure of the resulting membranes were characterized by scanning electron microscope, energy dispersive X-ray spectrometer, and powder X-ray diffraction. The $Cu_3(BTC)_2$/PVA membranes were then used in separating 50 wt % toluene/*n*-heptane mixtures through pervaporation. The effects of PVA concentration, $Cu_3(BTC)_2$ loading, feed composition, and operating temperature on membrane performances were explored. The results indicate that, compared with pristine PVA membrane, the separation factor and permeate flux of optimized $Cu_3(BTC)_2$/PVA membranes are improved from 8.9 and 14 g/(m^2 h) to 17.9 and 133 g/(m^2 h), respectively. A speculation of the transport process of permeating components in the selective layer of $Cu_3(BTC)_2$/PVA membrane was proposed. Enhanced affinity between toluene and the membrane through incorporating $Cu_3(BTC)_2$ particles plays a key role in improving separation performances.

Pervaporation is a potentially attractive process for the separation of dimethyl carbonate (DMC)/methanol mixture. Li. et al. [245] have studied the cross-linking density, swelling degree and pervaporation performance of the DMC-selective hydrophobic nano-silica/polydimethylsiloxane membrane by varying the membrane preparation parameters including cross-linking agent concentration, nano-silica loading, solvent concentration, operating time and temperature. The structural morphology, thermal stability and mechanical properties of the membrane were also investigated. Their results indicate that the thermal and mechanical stability of membrane can be enhanced by the well distributed silica incorporation, and the membrane containing 15 wt % hydrophobic nano-silica has the highest separation factor of 3.97 with a permeation flux of 702 g/(m^2 h) for 30 wt % DMC at the feed temperature of 40°C. Additionally, the absorption and diffusion selectivity of the membrane were measured and discussed via the solution–diffusion mechanism. The very low calculated diffusion selectivity of the membrane indicates that diffusion is the limiting factor in pervaporation performance.

New composite membrane with a poly(γ-benzyl-L-glutamate) (PBG) separation layer on a micro-porous poly(amide-imide) support was prepared and tested as a pervaporation membrane for selective separation of toluene/*n*-heptane mixtures [246]. Composite membrane was formed by coating a solution of PBG in chloroform on the top surface of micro-porous PAI-SO_2 support. PBG layer fulfils indispensable requirement for the preferential absorption of the most penetrating component (toluene) of a separated mixture.

The structure of the membrane was studied in detail. X-ray diffraction, electron microscopy and IR spectroscopy measurements were performed. Crystalline regions formed by backbones of macromolecules are low permeable barriers to penetrants, but also limiters for swelling of high permeable polymer phase. The knowledge of the material structure in dependence of conditions of its preparation enabled to adapt the process of the membrane formation to reach the efficiency of separating aromatic/aliphatic compounds.

The separation ability was tested by pervaporation measurements for different toluene concentrations in binary mixtures toluene/*n*-heptane. Diffusion coefficients, fluxes and separation factors were evaluated. It was shown that conformation of α-helix PBG molecules in the coating layer of the composite membrane does not change in the pervaporation. The membrane retains its integrity and can be used in subsequent cycles of a process.

The obtained results during this study could be used for the design of highly efficient PBG/PAI-SO_2 composite membranes for separation of mixtures aromatic and aliphatic hydrocarbons by pervaporation.

The separation of aromatic/aliphatic mixtures is significant in chemical industry. Pervaporation has attracted increasing attention due to its low energy consumption and environmentally friendly process. However, the formation of membranes mostly use toxic organic solvents, whereas in this study water was used instead to prepare membrane in separating aromatic/aliphatic mixtures. W3000 were dispersed in water to form micelles, which were then deposited onto the surface of the tubular porous ceramic substrate through the negative pressure-driven assembly method. The emulsion and composite membranes were characterized by TEM, SEM and FTIR. The crosslinking density of W3000 was also tested using NMR XLD analyzing system. The membranes were used for separating aromatic/aliphatic mixtures through pervaporation. The results showed that compared with the membrane prepared by DMF, the pervaporation separation index (PSI) of the composite membrane formed using water as solvent was 3.3-fold increased for separating toluene/*n*-heptane mixtures. The as-prepared composite membranes showed a better comprehensive pervaporation performance. This facile strategy may have a potential in the application of pervaporation for separating aromatic/aliphatic mixtures in industry [247].

3.7.2. Facilitation of Etherification

The present invention is an improved process for separating alcohols from ethers and/or hydrocarbon raffinate in an etherification process. The excess alcohol reactant, which forms azeotrope mixtures with the product ethers and C_4-C_7 raffinate, is removed by passing the liquid azeotrope mixture over a pervaporation membrane which effectively breaks the azeotrope and permeates the alcohol with high flux and high selectivity. In a typical etherification process, one or more pervaporation membrane units can be located ahead of the ether/raffinate distillation step, in conjunction with the distillation step with a liquid side draw, after the distillation step, or a combination of any of the above. The present invention also provides an improved process for separating alcohols from ethers and/or hydrocarbon raffinate in an ether decomposition process for the production of high purity iso-alkene products. In this embodiment, one or more pervaporation membranes are used to recover alcohols from the decomposition product stream [248].

The present invention is an improved process for separating alcohols from ethers and/or hydrocarbon raffinate in an etherification process [249]. The excess alcohol reactant which forms azeotrope mixtures with the product ethers and C_4-C_7 raffinate, is removed by passing the liquid azeotrope mixture over a pervaporation membrane which effectively breaks the azeotrope and permeates the alcohol with high flux and high selectivity. In a typical etherification process, one or more pervaporation membrane units can be located ahead of the ether/raffinate distillation step, in conjunction with the distillation step with a liquid side draw, after the distillation step, or a combination of any of the above. The present invention also provides an improved process for separating alcohols from ethers and/or hydrocarbon raffinate in an ether decomposition process for the production of high purity iso-alkene products. In this embodiment, one or more pervaporation membranes are used to recover alcohols from the decomposition product stream.

The etherification of *tert*-amyl alcohol with ethanol was carried out in a reactive distillation column inserted by a zeolite NaA membrane tube [250]. Experimental tests were carried out in both of a pervaporation module and a reactive distillation column. Under suitable conditions, the pervaporation tests have shown higher than 99.9% H_2O mole fraction in the permeate. The design by the residue curve maps has shown the alleviation of azeotropes of H_2O-reaction components mixtures under pervaporation. The experimental study at standard conditions has shown a gain of 10% in *tert*-amyl ethyl ether (TAEE) yield when the zeolite membrane tube was inserted inside the distillation column. Further improvements in TAEE yield were realized when the feed location was separated and the time factor or the reflux ratios was increased.

The combined process of pervaporation with reactive distillation was studied for the production of ethyl *tert*-butyl ether (ETBE) from ethanol (EtOH) and *tert*-butyl alcohol (TBA) on an ion-exchange catalyst [251]. An apparatus consisting of a stirred batch reactor, a distillation column, and a pervaporation membrane was used to test this technique. The permeation flux and selectivity of water in the membrane were investigated in water-alcohols system. The etherification was performed in the liquid phase by using a batch reactor. The reactive distillation was examined with and without pervaporation at the boiling point of the reactant mixture under atmospheric pressure. When pervaporation was not conducted, two layers formed in the top products because of a higher concentration of water. However, these phenomena were not observed when pervaporation was added. It was revealed that pervapovation might be effective for removing water from the bottom and that a higher fraction of ETBE could be obtained as a top product.

Asymmetric membranes were prepared by first forming multi-walled carbon nanotube-buckypaper (MWCNT-BP) structures and later coating the structures with a thin layer of polyvinyl alcohol (PVA) to form novel MWCNT-BP/PVA asymmetric membranes. The resultant asymmetric membranes were used for the dehydration of a multi-component reaction mixture obtained from etherification through pervaporation. The asymmetric membranes exhibited improved mechanical properties relative to those of a pure PVA membrane. When the purified MWCNT-BP/PVA asymmetric membranes were applied in pervaporation, they exhibited significant two- and four-fold enhancements of the permeation flux and separation factor, respectively, compared to those of a pure PVA membrane. This effect was observed due to the hydrophilic group on the oxidized MWCNTs and the nanochannels of the pre-selective layer, which favor the permeation of water molecules [252].

Novel approach to obtain high yield to poly-*tert*-butylglycerolethers by glycerol etherification reaction with tert-butyl alcohol (TBA) is proposed by Frusteri et al. [253]. The limit of this reaction is the formation of water which inhibits the formation of poly-ethers potentially usable in blend with conventional diesel for transportation. The results here reported demonstrate that the use of a water permselective membrane offers the possibility to shift the equilibrium towards the formation of poly-ethers since the water formed during reaction is continuously and selectively removed from the reaction medium by the recirculation of the gas phase. Using a proper catalyst and by optimizing the reaction conditions, in a single experiment, a total glycerol conversion can be reached with a yield to poly-ethers close to 70% which represents a data never reached up to now using TBA as reactant. The approach here proposed could open new opportunity for all catalytic reactions affected by water formation.

Etherification of glycerol with (*bio*)butanol was investigated to obtain glycerol ethers usable as oxygenated additives for diesel fuel [254]. Experiments were carried out using a batch reactor assisted by a tubular water-permselective membrane. A-15 resin was used as solid acid catalyst at a reaction temperature ranging from 70 to 160°C in view of maximizing the glycerol conversion. The water formed during the reaction was continuously removed by recirculation the gas phase through the membrane and this allowed to enhance reactions kinetics reaching both a total glycerol conversion and the formation of *poly*-substituted ethers not obtainable without the membrane. The alcohol dehydration and self-etherification of glycerol or alkyl alcohols give rise to the formation of several side products at moment hardly to identify.

Etherification of glycerol with (*bio*)butanol was investigated to obtain glycerol ethers usable as oxygenated additives for diesel fuel. Experiments were carried out using a batch reactor assisted by a tubular water-permselective membrane. A-15 resin was used as solid acid catalyst at a reaction temperature ranging from 70 to 160°C in view of maximizing the glycerol conversion. The water formed during the reaction was continuously removed by recirculation the gas phase through the membrane and this allowed to enhance reactions kinetics reaching both a total glycerol conversion and the formation of *poly*-substituted ethers not obtainable without the membrane. The alcohol dehydration and self-etherification of glycerol or alkyl alcohols give rise to the formation of several side products at moment hardly to identify [255].

The present invention is an improved process for separating alcohols from ethers and/or hydrocarbon raffinate in an etherification process. The excess alcohol reactant which forms azeotrope mixtures with the product ethers and C4-C7 raffinate, is removed by passing the liquid azeotrope mixture over a pervaporation membrane which effectively breaks the azeotrope and permeates the alcohol with high flux and high selectivity. In a typical etherification process, one or more pervaporation membrane units can be located ahead of the ether/raffinate distillation step, in conjunction with the distillation step with a liquid side draw, after the distillation step, or a combination of any of the above. The present invention also provides an improved process for separating alcohols from ethers and/or hydrocarbon raffinate in an ether decomposition process for the production of high purity iso-alkene products. In this embodiment, one or more pervaporation membranes are used to recover alcohols from the decomposition product stream [256].

Pervaporation is a membrane separation process vastly used for purification in chemical and allied industries. Esterification reaction can be intensified and enhanced by coupling with pervaporation reactor (PVR). Polyvinyl alcohol (PVA)/polyether sulfone (PES) composite membrane was used for the pervaporation coupled esterification reaction study. Esterification of butyric acid with n-propanol was taken as a model reaction for the study and to test the performance of pervaporation reactor. Catalyst p-toluene sulfonic acid was used for the esterification reaction. The effects of various reaction parameters on conversion of butyric acid such as reaction temperature, initial molar ratio of n-propanol to butyric acid, catalyst loading and reaction time were studied. Experimental results show that the increase of temperature, initial molar ratio, and catalyst concentration enhance the conversion of butyric acid considerably. The highest conversion of 96.41% was obtained at temperature 353 K, molar ratio of 2 and catalyst loading of 2.5%w/w at reaction time of 420 minutes. PVA/PES membrane used in the experiments shows the good activity and hydrophilicity and plays a vital role for enhancing the conversion by selectively removing water. Pervaporation coupled esterification shows the better choice over the conventional route for the production of esters. This technique is environment friendly and energy intensified approach as it reduces pollution and energy requirement [257].

REFERENCES

[1] Binding, R. C., Lee, R. J., Jennings, J. F., Mertin, E. C. (1961) Separation of liquid mixtures by permeation. *Industrial Engineering Chemistry*, 53: 45-54.

[2] Choo, C. Y. (1962) Membrane permeation in advances in petroleum chemistry and refining, Chapter 2. In: Mcketta, J. (ed.) *Advances in petroleum chemistry and refining,* vol IV. Interscience, New York.

[3] Kahlenberg, L. (1906) On the nature of the process of osmosis and osmotic pressure with observations concerning dialysis., *Journal of Physical Chemistry*, 10: 141-209.

[4] Kammermeyer, K., Hagerbaumer, D. H. (1955) Membrane separation in the liquid phase. *AIChE Journal*, 1, 215-219].

[5] Heisler, E. G., Hunter, A. S., Siciliano, J., Treadway, R. M. (1956) Solute and temperature effects in the pervaporation of aqueous alcoholic solutions. *Science,* 124: 77-79.

[6] Michaels, A. S., Baddour, R. F., Bixler, H. J., Choo, C. Y. (1962) Conditioned polyethylene as a permselective membrane. Separation of isomeric xylenes. *Industrial & Engineering Chemistry Process Design and Development*, 1, 14-25.

[7] Strazik, W. F., Perry, E. (1972) *Process for the separation of styrene from ethyl benzene.* US Patent 3,776,970A.

[8] Perry E. (1975) *Membrane separation of organics from aqueous solutions.* US Patents 4,218,312A, and many others.

[9] Aptel, P., Challard, N., Cuny, J., Neel, J. (1976) Application of the pervaporation process to separate azeotropic mixtures, *Journal of Membrane Science*, 1, 271-287.

[10] Neel, J., Nguyen, Q. T., Clement, R., Le Blanc, L. (1983) Fractionation of a binary liquid mixture by continuous pervaporation, *Journal of Membrane Science*, 15, 43-62.

[11] Cabasso, I., Jagur-Grodzinski, J., Vofsi, D. (1974) Polymeric alloys of polyphosponates and acetyl cellulose. Absorption and diffusion of benzene and cyclohexane. *Journal of Applied Polymer Science*, 18, 2117-2136.

[12] Cabasso, I., Jagur-Grodzinski, J., Vofsi, D. (1974) A study of permeation of organic solvents through polymeric membranes based on polymeric alloys of polyphosphonates and acetyl cellulose. II.

Separation of benzene, cyclohexene, and cyclohexane. *Journal of Applied Polymer Science*, 18, 2137-2147.

[13] Sikonia, J. G., McCandless, F. P. (1978) Separation of isomeric xylenes by permeation through modified plastic films. *Journal of Membrane Science*, 4, 229-241.

[14] Uragami, T. (2006) Polymer membranes for separation of organic liquid mixtures. In *Materials Science of Gas and Vapor Separation,* Y. Yanpoliski, I. Pinnau, B. D. Freeman (eds.). John Wiley & Sons, Chichester, pp. 355-372.

[15] Uragami, T. (2010) Selective membranes for purification and separation of organic liquid mixtures. In *Comprehensive Membrane Science and Engineering,* Volume 2, E. Drioli, L. Giorno (eds.). Elsevier, Amsterdam, pp. 273-324.

[16] Uragami, T. (2015) Pervaporation Membranes. In *Encyclopedia of Polymeric Nanomaterials*, Volume 2, S. Kobayashi, K. Müellen (eds.). Springer, Berlin, Heidelberg, pp. 156-158.

[17] Uragami, T. (2017) Pervaporation. In *Science and Technology of Separation Membranes*, Volume II, John Wiley & Sons, Chichester, pp. 413-417.

[18] Uragami, T. (2017) Selective membranes for purification and separation of organic liquid mixtures. In *Comprehensive Membrane Science and Engineering,* Volume 2, E. Drioli, L. Giornopp, E. Fontananova (eds.). Elsevier, Amsterdam, pp. 256-331.

[19] Hamatani, T., Yamada, S. (1980) Permeation separation characteristics aqueous dimethylamine solutions through cellophane membrane. *Kobunshi Ronbunshu*, 37, 65-71 (Japanease).

[20] Uragami, T. (2017) Section 2.10, Selective membranes for purification and separation of organic liquid mixtures, In *Comprehensive Mem-brane Science and Engineering 2nd Edition*, Volume 2, E. Drioli, Giorno, L., Fontananvoa, E. ‘eds. Elsevier, Amsterdam, Boston, Hidelberg, London, New York, Oxford, Paris, San Diego, San Francisco, Singapore, Sydney, Tokyo, pp. 264.

[21] Ihm, C. D., Ihm, S. K. (1995) Pervaporation of water-ethanol mixtures through sulfonated polystyrene membranes prepared by plasma graft-polymerization. *Journal of Membrane Science*, 98: 89-96.

[22] Yeom, C. K., Jegal, J. G., K. H. Lee, K. H. (1996) Characterization of relaxation phenomena and permeation behaviors in sodium alginate membrane during pervaporation separation of ethanol–water mixture. *Journal of Applied Polymer Sci*ence, 62: 1561-1576.

[23] Jegal, J., Kew-Ho Lee, H. (1996) Development of polyion complex membranes for the separation of water–alcohol mixtures. III. Preparation of polyion complex membranes based on the -carrageenan for the pervaporation separation of water–ethanol. *Journal of Applied Polymer Science*, 60, 1177-1183.

[24] Richau, K., Schwarz, H. H., Apostel, R., Paul, D. (1996) Dehydration of organics by pervaporation with polyelectrolyte complex membranes: some considerations concerning the separation mechanism. *Journal of Membrane Science*, 113, 31-41.

[25] Nam, S. Y., Young MooLee, Y. M. (1997) Pervaporation and properties of chitosan-poly(acrylic acid) complex membranes. *Journal of Membrane Science*, 135: 161-171.

[26] Iwatsubo, T., Kusumocahyo, S. P., Shinbo, T. (2002) Water/ethanol pervaporation performance of asymmetric polyelectrolyte complex membrane constructed by the diffusion of poly(acrylic acid) in chitosan membrane. *Journal of Applied Polymer Science*, 86, 265-271.

[27] Uragami, T. Waketa, D. (2010) Dehydration of an azeotrope of ethanol/water by sodium carboxymethylcellulose membranes cross-linked with organic or inorganic cross-linker. *eXPRESS Polymer Letters*, 4, 681-691.

[28] Uragami, T., Banno, M., Miyata, T. (2015) Dehydration of an ethanol/water azeotrope through alginate-DNA membranes cross-linked with metal ions by pervaporation. *Carbohydrate Polymers*, 134, 38-45.

[29] Jegal, J., Kew-Ho Lee, H. (1996) Development of polyion complex membranes for the separation of water–alcohol mixtures. III.

Preparation of polyion complex membranes based on the κ-carrageenan for the pervaporation separation of water–ethanol. *Journal of Applied Polymer Science*, 60, 1177-1183.

[30] Richau, K., Schwarz, H. H., Apostel, R., Paul, D. (1996) Dehydration of organics by pervaporation with polyelectrolyte complex membranes: some considerations concerning the separation mechanism. *Journal of Membrane Science*, 113: 31-41.

[31] Nam, S. Y., Lee, Y. M. (1997) Pervaporation and properties of chitosan-poly(acrylic acid) complex membranes. *Journal of Membrane Science*, 135: 161-171.

[32] Iwatsubo, T., Kusumocahyo, S. P., Shinbo, T. (2002) Water/ethanol pervaporation performance of asymmetric polyelectrolyte complex membrane constructed by the diffusion of poly(acrylic acid) in chitosan membrane. *Journal of Applied Polymer Science*, 86: 265-271.

[33] Uragami, T., Yamamoto, S. Miyata, T. (1999) Preparation and characteristics of polyioncomplex crosslinked chitosan membranes. *Network Polymer*, 20, 203-208.

[34] Uragami, T. Yamamoto, S. Miyata, T. (2003) Dehydration from alcohols by polyion complex cross-linked chitosan composite membranes during evapomeatuion. *Biomacromolecules*, 4: 137-144.

[35] Lai, J. Y. Chu, Y. H. Huang, S. Yin, Y. L. (1994) Separation of water–alcohol mixtures by pervaporation through asymmetric nylon 4 membrane. *Journal of Applied Polymer Science*, 53: 999-1009.

[36] Chen, S. H. Lai, J. Y. (1995) Chemical grafted 4-vinylpyridine onto polycarbonate membranes for pervaporation. *Journal of Applied Polymer Science*, 55: 1353-1359.

[37] Lee, K. R., Lai, J. Y. (1995) Pervaporation of aqueous alcohol mixtures through a membrane prepared by grafting of polar monomer onto nylon 4. *Journal of Applied. Polymer Science*, 57: 961-968.

[38] Sun, Y. M., Huang, T. L. (1996) Pervaporation of ethanol-water mixtures through temperature-sensitive poly (vinyl alcohol-*g*-*N*-isopropylacrylamide) membranes. *Journal of Membrane Science*, 110, 211-218.

[39] Buyanov, A. L., Revel'skaya, L. G., Kuznetzov, Y. P., Shestakova, A. S. (1998) Cellulose–poly(acrylamide or acrylic acid) interpenetrating polymer network membranes for the pervaporation of water–ethanol mixtures. *Journal of Applied Polymer Science*, 69, 761-769.

[40] Shien, J. J. Huang, R. Y. M. (1997) Pervaporation with chitosan membranes II. Blend membranes of chitosan and polyacrylic acid and comparison of homogeneous and composite membrane based on polyelectrolyte complexes of chitosan and polyacrylic acid for the separation of ethanol-water mixtures. *Journal of Membrane Science*, 127: 185-202.

[41] Zhang, L. N. Guo, J. Zhou, J. P. Yang, G. Du, Y. M. (2000) Blend membranes from carboxymethylated chitosan/alginate in aqueous solution. *Journal of Applied Polymer Science*, 77, 610-616.

[42] Uragami, T. Matsuda, Okuno, H. Miyata, T. (1994) Structure of chemically modified chitosan membranes and their characteristics of permeation and separation of aqueous ethanol solutions. *Journal of Membrane Science,* 88, 243-251.

[43] Uragami, T. Okazaki, K. Matsugi, H. Miyata, T. (2002) Structure and permeation characteristics of an aqueous ethanol solution of organic-inorganic hybrid membranes composed of poly (vinyl alcohol) and tetraethoxysilane. *Macromolecules*, 35, 9156-9163.

[44] Uragami, T. Katayama, T. Miyata, T. Tamura, H. Shiwaiwa, T. Higuchi, A. (2004) Dehydration of an ethanol/water azeotrope by novel organic-inorganic hybrid membranes based on quatemized chitosan and tetraethoxysilane. *Biomacromolecules*, 5, 2116-2121.

[45] Uragami, T., Matsugi, H., T. Miyata, T. (2005) Pervaporation characteristics of organic-inorganic hybrid membranes composed of poly (vinyl alcohol-co-aciylic acid) and tetraethoxysilane for water/ethanol separation. *Macromolecule*, 38, 8440-8446.

[46] Uragami, T. Yanagisawa, S. Miyata, T. (2007) Water/ethanol selectivity new organic-inorganic hybrid membranes fabricated from poly(vinyl alcohol) and an olygosilane, *Macromolecular Chemistry and Phys*ics 208, 756-764.

[47] Lee, K. R., Chen R. Y., Lai, J. Y. (1992) Plasma deposition of vinyl acetate onto Nylon-4 membrane for pervaporation and evapomeation separation of aqueous alcohol mixtures. *Journal of Membrane Science*, 75, 171-180.

[48] Lai, J. Y., Chen, R. Y., Lee, K., R. (1993) Polyvinyl alcohol γ-ray grafted nylon 4 membrane for pervaporation and evapomeation. *Separation Science Technology*, 28, 1437-1452.

[49] Xu, Y. X., Chen, C. X., Zhang, P., X. Sun, B. H., Li, J. D. (2006) Pervaporation properties of polyimide membranes for separation of ethanol + water mixtures. *Journal of Chemical Engineering Data*, 51, 1841-1845.

[50] Huang, Z., Shi, Y., Wen, R., Guo, Y. H., Su, J. F., Matsuura, T. (2006) Multilayer poly (vinyl alcohol) –zeolite 4A composite membranes for ethanol dehydration by means of pervaporation. *Separation Purification Technology*, 51, 126-136.

[51] Fan, S. C., Wang, Y. C., Li, C. L., Lee, K. R., Liaw, D. J., Lai, J. Y. (2003) Permselectivities of 2,2′-dimethyl-4,4′-bis(aminophenoxyl) biphenyl diphenyl methane–based aromatic polyamide membranes for aqueous alcohol mixtures in pervaporation and evapomeation. *Journal of Applied Polymer Science*, 88, 2688-2697.

[52] Li, C. L., Lee, K. R. (2006) Dehydration of ethanol/water mixtures by pervaporation using soluble polyimide membranes. *Polymer International*, 55, 505-512.

[53] Zhang, Q. G., Liu, Q. L., Jiang, Z. Y., Chen, Y. (2007) Anti-trade-off in dehydration of ethanol by novel PVA/APTEOS hybrid membranes. *Journal of Membrane Science*, 287, 237-245.

[54] Gimenes, M. L., Liu, L., Feng, X. S. (2007) Sericin/poly (vinyl alcohol) blend membranes for pervaporation separation of ethanol/water mixtures. *Journal of Membrane Science*, 295, 71-79.

[55] Namboodiri, V. V., Vane, L. M. (2007) High permeability membranes for the dehydration of low water content ethanol by pervaporation. *Journal of Membrane Science*, 306, 209-215.

[56] Teli, S. B., Gokavi, G. S., Aminabhavi, T. M. (2007) Novel sodium alginate-poly(N-isopropylacrylamide) semi-interpenetrating polymer

network membranes for pervaporation separation of water+ethanol mixtures. *Separation and Purification. Technology*, 56, 150-157.

[57] Shih, C. Y., Chen, S. H., Liou, R. M., Lai, J. Y., Chang, J. S. (2007) Pervaporation separation of water/ethanol mixture by poly (phenylene oxide) and sulfonated poly(phenylene oxide) membranes. *Journal of Applied Polymer Science*, 105, 1566-1574.

[58] Chen, S. H., Liou, R. M., Shih, C. Y., Lai, C. L., Lai, J. Y. (2007) Pervaporation separation of water/ethanol mixture using asymmetric ion-exchange membranes containing aluminum ions. *Journal of Applied Polymer Science*, 106, 1412-1420.

[59] Kim, S. G., Lee, K. S., Lee, K. H. (2006) Polyelectrolyte complex membranes based on two anionic polysaccharides composed of sodium alginate and carrageenan: The effect of annealing on the separation of methanol/water mixtures. *Journal of Applied Polymer Science*, 102, 5781-5788.

[60] Lee, C., Hong, W. H. (1997) Influence of different degrees of hydrolysis of poly (vinyl alcohol) membrane on transport properties in pervaporation of IPA/water mixture. *Journal of Membrane Science*, 135, 187-193.

[61] Bhat, N. M., Wavhal, D. S. (2000)Preparation of cellulose triacetate pervaporation membrane by ammonia plasma treatment. *Journal of Applied Polymer Science*, 76, 258-265.

[62] Bhat, N. V., Wavhal, D. S. (2000) Characterization of plasma-polymerized thiophene onto cellulose acetate membrane and Iits application to pervaporation. *Separation Science Techno*logy, 35, 227-242.

[63] Toti, U. S. Aminabhavi, T. M. (2002) *Journal of Applied Polymer Science*, 85, 2014-2024. Pervaporation separation of water-isopropyl alcohol mixtures with blend membranes of sodium alginate and poly(acrylamide)-grafted guar gum.

[64] Liu, Y. L., Su, Y. H., Lee, K. R., Lai, J. Y. (2005) Crosslinked organic–inorganic hybrid chitosan membranes for pervaporation dehydration of isopropanol–water mixtures with a long-term stability. *Journal of Membrane Science*, 251, 233-238.

[65] Qiao, X. Y., Chung, T. S. (2006) Diamine modification of P84 polyimide membranes for pervaporation dehydration of isopropanol. *AIChE Journal*, 52, 3462-3472.

[66] Xiao, S. D., Huang, R. Y. M., Feng, X. S.(2006) Preparation and properties of trimesoyl chloride crosslinked poly(vinyl alcohol) membranes for pervaporation dehydration of isopropanol *Journal of Membrane Science*, 286, 245-254.

[67] Liu, Y. L., Yu, C. H., Lee, K. R., Lai, J. Y. (2007) Chitosan/ poly(tetrafluoroethylene) composite membranes using in pervaporation dehydration processes. *Journal of Membrane Science*, 287, 230-236.

[68] Liu, R. X., Qiao, X. Y., Chung, T. S. (2007) Dual-layer P84/polyethersulfone hollow fibers for pervaporation dehydration of isopropanol. *Journal of Membrane Science*, 294, 103-114.

[69] Choudhari, S. K., Kittur, A. A., Kulkarni, S. S., Kariduragan, M. Y. (2007) Development of novel blocked diisocyanate crosslinked chitosan membranes for pervaporation separation of water–isopropanol mixtures. *Journal of membrane Science*, 302, 197-206.

[70] Veerapur, R. S. Gudasi, K. B. Aminabhavi, T. M. (2007) Pervaporation dehydration of isopropanol using blend membranes of chitosan and hydroxypropyl cellulose. *Journal of Memrane Science*, 304, 102-111.

[71] Teli, S. B., Gokavi, G. S., Sairam, M., Aminabhavi, T. M. (2007) Mixed matrix membranes of poly (vinyl alcohol) loaded with phosphomolybdic heteropolyacid for the pervaporation separation of water–isopropanol mixtures. *Colloids and Surfaces A: Physicochemical and Engineering*, 301, 55-62.

[72] Shen, J. N., Wu, L. G., Qiu, J. H., Gao, C. J. (2007) Pervaporation separation of water/isopropanol mixtures through crosslinked carboxymethyl chitosan/polysulfone hollow-fiber composite membranes. *Journal of Applied Polymer Science*, 103, 1959-1965.

[73] Jiang, L. Y., Chung, T. S., Rajagopalan, R. (2008) Dehydration of alcohols by pervaporation through polyimide Matrimid® asymmetric

hollow fibers with various modifications. *Chemical Engineering Science*, 63, 204-216.

[74] Slater, C. S. Schurmann, T. Zimarowski, A., MacMillian, J. (2006) Separation of diactenone alcohol-water mixtures by membrane pervaporation. *Separation Science and Technology*, 41, 2733-2753.

[75] Devi, D. A. Smitha, B. Sridhar, S. Jawalkar, S. S. Aminabhavi, T. M. (2007) Novel sodium alginate/polyethyleneimine polyion complex membranes for pervaporation dehydration at the azeotropic composition of various alcohols. *Journal of Chemical Technology and Biotech*nology, 82, 993-1003.

[76] Kim, S. G., Lee, K. S., Lee, K. H. (2007) Pervaporation 0f sodium alginate/chitosan polyelectrolete complex composite membranes for the separation of water/alcohol mixtures: Characterization of the permeation behavior with molecular modeling techniques. *Journal of Applied Polymer Science*, 103, 2634-2641.

[77] Hu, C. L., Guo, R. L., Li, B., Ma, X. C., Wu, H., Jiang, Z. Y. (2007) Development of novel mordenite-filled chitosan–poly (acrylic acid) polyelectrolyte complex membranes for pervaporation dehydration of ethylene glycol aqueous solution. *Journal of Membrane Science*, 293, 142-150.

[78] Rao, P. S., Sridhar, S., Wey, M. Y., Krishnaiah, A. (2007) Pervaporative separation of ethylene glycol/water mixtures by using cross-linked chitosan membranes. *Industrial Engineering Chemistry Research*, 46, 2155-2163.

[79] Guo, R. L., Hu, C. L., Li, B., Jiang, Z. Y.(2007) Pervaporation separation of ethylene glycol/water mixtures through surface crosslinked PVA membranes: Coupling effect and separation performance analysis. *Journal of Membrane Science,* 289, 191-198.

[80] Shi, G. M., Yang, T., Chung, T. S. (2012) Polybenzimidazole (PBI)/zeolitic imidazolate frameworks (ZIF-8) mixed matrix membranes for pervaporation dehydration of alcohols. *Journal of Membrane Science*, 415-416, 577-586.

[81] Kuila, S. B., Ray, S. K. (2014) Dehydration of dioxane by pervaporation using filled blend membranes of polyvinyl alcohol and sodium alginate. *Carbohydrate Polymers*, 101, 1154-1165.

[82] Xia, L. L., Li, C. L., Wang, Y. (2016) In-situ crosslinked PVA/organosilica hybrid membranes for pervaporation separations. *Journal of Membrane Science,* 498, 263-275.

[83] Miyata, T. Takagi, T. Kadota, T. Uragami, T. (1995) Characteristics of permeation and separation for aqueous ethanol solutions through methyl methacrylate-dimethyl siloxane graft copolymer membranes. *Macromolecular Chemistry and Physics*. 196, 1211-1220.

[84] Miyata, T. Takagi, T. Uragami, T. (1996) Microphase separation in graft copolymer membranes with pendant oligodimethylsiloxanes and their permselectivity for aqueous ethanol solutions, *Macromolecules*, 29, 7787- 7794.

[85] Miyata, T. Obata, S. Uragami, T. (1999) Morphological Effects of Microphase Separation on The Permselectivity for Aqueous Ethanol Solutions of Block and Graft Copolymer Membranes Containing Poly(dimethyl siloxane). *Macromolecules*, 32, 3712- 3720.

[86] Miyata, T. Obata, S. Uragami., T. (1999) Annealing effect of microphase-separated membranes containing poly(dimethyl siloxane) on their permselectivity for aqueous ethanol solutions. *Macromolecules*, 32, 8465- 8475.

[87] Ishihara, K. Nagase, Y. Matsui, K. (1986) Pervaporation of alcohol/water mixtures through poly[1-(trimethylsilyl)-1-propyne] membrane. *Makromolecular Rapid Commun*ications, 7, 43- 46.

[88] Masuda, T., Tang, B. Z., Higashimura, T. (1986) Ethanol-water separation by pervaporation through substututed polyacetylne membranes. *Polymer Journal,* 18, 565- 567.

[89] Uragami, T., Doi, T., Miyata, T. (1999) Control of permselectivity with surface modifications of poly[1-(trimethylsilyl)-1-propyne] membranes. *International Journal of Adhesion & Adhesive*, 19, 405-409.

[90] Uragami, T., Doi, T., Miyata T. Pervaporation Property of Surface Modified Poly[1-(trimethylsilyl)-1propyne] Membranes, In

Membrane Formation and Modification, ACS Symposium Series 744, Pinnau, I., Freeman, B. D. Eds. American Chemical Society: Washington, DC, 2000, pp 263- 279.

[91] Volkov, V. V., Fadeev, A. G., Khotimsky, V. S., Litvinova, E. G., Selinskaya, Y. A., McMillan, J. D., Kelley, S. S. (2004) Effects of synthesis conditions on the pervaporation properties of poly[1-(trimethylsilyl) -1-propyne] useful for membrane bioreactors. *Journal of Applied Polymer Science*, 91, 2271-2277.

[92] Wasewar, K. L., Pangarkar, V. G. (2006) Intensification of recovery of ethanol from fermentation broth using pervaporation: Economical evaluation, *Chemical and Biochemical Engineering Quarterly*, 20, 135-145.

[93] Shi, S. P., Du, Z. J., Ye, H., Zhang, C., Li, H. Q. (2006) A Novel Carbon Black/Polydimethylsiloxane Composite Membrane with High Flux for the Separation of Ethanol from Water by Pervaporation. *Polymer Journal*, 38, 949-955.

[94] Bowen, T. C., Meier, R. G., Vane, L. M. (2007) Stability of MFI zeolite-filled PDMS membranes during pervaporative ethanol recovery from aqueous mixtures containing acetic acid. *Journal of Membrane Science*, 298, 117-125

[95] Xiangli, F. J. Chen, Y. W. Jin, W. Q. Xu, N. P. (2007) Polydimethylsiloxane (PDMS)/ceramic composite membrane with high flux for pervaporation of ethanol−water mixtures. *Industrial & Engineering Chemistry* Research, 46, 2224-2230.

[96] Lewandowska, M., Kujawski, W. (2007) Ethanol production from lactose in a fermentation/pervaporation system. *Journal of Food. Engineering*, 79, 430-437.

[97] Ei-Zanati, E., Abdel-Hakim, E., El-Ardi, O., Fahmy, M. (2006) Modeling and simulation of butanol separation from aqueous solutions using pervaporation. *Journal of Membrane Science,* 280, 278-283.

[98] Srinivasan, K., Palanivelu, K., Gopalakrishnan, A. N. (2007) Recovery of 1-butanol from a model pharmaceutical aqueous waste by pervaporation. *Chemical Engineering Science*, 62, 2905-2914.

[99] Zhuang, X., Chen, X., Su, Y., Luo, J., Feng, S., Zhou, H., Wan, Y. (2016) Surface modification of silicalite-1 with alkoxysilanes to improve the performance of PDMS/silicalite-1 pervaporation membranes: Preparation, characterization and modeling. *Journal of Membrane Science*, 499, 386-395.

[100] Wang, X., Chen, J., Fang, M., Wang, T., Yu, L., Li, J. (2016) ZIF-7/PDMS mixed matrix membranes for pervaporation recovery of butanol from aqueous solution. *Separation and Purification Technology*, 163, 39-47.

[101] Naik, P. V., S., Martens, J. A., Vankelecom, I. F. J. (2016) PDMS mixed matrix membranes containing hollow silicalite sphere for ethanol/water separation by pervaporation. *Journal of Membrane Science*, 502, 48-56.

[102] Gao, L., Alberto, M., Gorgojo, P., Szekely, G., Budd,P. M. (2017) High-flux PIM-1/PVDF thin film composite membranes for 1-butanol/water pervaporation. *Journal of Membrane Science*, 529, 207-214.

[103] Zhou, H., Zhang, J., Wan, Y., Jin, W. (2017) Fabrication of high silicalite-1 content filled PDMS thin composite pervaporation membrane for the separation of ethanol from aqueous solutions. *Journal of Membrane Science*, 524, 1-11.

[104] Singha, N. R., Parya, T. K., Ray, S. K. (2009) Dehydration of 1,4-dioxane by pervaporation using filled and crosslinked polyvinyl alcohol membrane. *Journal of Membrane Science*, 340, 35-44.

[105] Bhat, S. D., Aminabhavi, T. M. (2006) Novel sodium alginate–Na^+MMT hybrid composite membranes for pervaporation dehydration of isopropanol, 1,4-dioxane and tetrahydrofuran. *Separation. Purification Technology*, 51, 85-94.

[106] Bhat, S. D., Mallikarjuna, N. N., Aminabhavi, T. M. (2006) Microporous alumino-phosphate ($AlPO_4$-5) molecular sieve-loaded novel sodium alginate composite membranes for pervaporation dehydration of aqueous–organic mixtures near their azeotropic compositions. *Journal of Membrane Science*, 282, 473-483.

[107] Premakshi· H. G., K. Ramesh· K., M. Y. Kariduraganavar· M. Y. (2015) Modification of crosslinked chitosan membrane using NaY zeolite for pervaporation separation of water–isopropanol mixtures. *Chemical Engineering Research and Design*, 94, 32-43.

[108] Lia, Y. L., Hu, C. C., Lai, J. Y., Ying-LingLiu, Y. L. (2017) Crosslinked polybenzoxazine based membrane exhibiting in-situ self-promoted separation performance for pervaporation dehydration on isopropanol aqueous solutions. *Journal of Membrane Science*, 531, 10-15.

[109] Sairam, M. Naidu, B. V. K. Nataraj, S. K. Sreedhar,B. Aminabhavi, T. M. (2006) Poly(vinyl alcohol)-iron oxide nanocomposite membranes for pervaporation dehydration of isopropanol, 1,4-dioxane and tetrahydrofuran. *Journa of Membrane Science*, 283, 65-73.

[110] Smitha, B. Dhanuja, G. Sridhar, S. (2006) Dehydration of 1,4-dioxane by pervaporation using modified blend membranes of chitosan and nylon 66. *Carbohydrate Polymers*, 66, 463-472.

[111] Ray, S. Ray, S. K. (2007) Synthesis of highly selective copolymer membranes and their application for the dehydration of tetrahydrofuran by pervaporation. *Journal of Applied Polymer Science*, 103, 728-737.

[112] Krishna Rao, K. S. V., Subha, M. C. S., Sairam, M., Mallikarjuna, N. N., T. M. Aminabhavi、T. M. (2006) Blend membranes of chitosan and poly(vinyl alcohol) in pervaporation dehydration of isopropanol and tetrahydrofuran. *Journal of Applied Polymer Science*, 103, 1918-1926.

[113] [113] Rao, P. S. Sridhar, S. Wey, M. Y. Krishnaiah, A. (2007) Pervaporation performance and Transport phenomenon of PVA blend membranes for the separation of THF/water azeotropic mixtures. *Polymer Bulletin*, 59, 289-298.

[114] Reddy, K. M., Babu, V. R.,Sariam, M., Aminabhavi, T. M. (2007) Sodium alginate-TiO₂ mixed matrix membranes for pervaporation dehydration of tetrahydrofuran and isopropanol. *Designed Monomers & Polym*ers. 10, 297-309.

[115] Bhat, M., S. D., Aminabhavi, T. M. (2007) Zeolite K-LTL-loaded sodium alginate mixed matrix membranes for pervaporation dehydration of aqueous–organic mixtures. *Journal of Membrane Science*, 306, 173-185.

[116] Solak, E. K. Sanli, O. (2006) Separation characteristics of dimethylformamide/water mixtures through alginate membranes by pervaporation, vapor permeation and vapor permeation with temperature difference methods. *Separation Science Technology*, 41, 627-646.

[117] Lai, J. Y., Yin, Y. L., Lee, K. R. (1995) Chemically modified poly(4-methyl-1-pentene) membrane for pervaporation separation of acetic acid-water mixtures. *Polymer Journal*, 27, 813- 818.

[118] Asman, G., Sanli, O. (2003) Characteristics of permeation and separation for acetic acid–water mixtures through poly(vinyl alcohol) membranes modified with poly(acrylic acid). *Separation Science and. Technology*, 38, 1963-1980.

[119] Isiklan, N., Sanli, O. (2004) Permeation and separation characteristics of acetic acid/water mixtures through poly (vinyl alcohol-*g*-itaconic acid) membranes by pervaporation, evapomeation, and temperature-difference evapomeation. *Journal of Applied Polymer Science*, 93, 2322-2333.

[120] Asman, G., Sanli, O. (2006) Separation characteristics of acetic acid–water mixtures using poly (vinyl alcohol-*g*-4-vinyl pyridine) membranes by pervaporation and temperature difference evapomeation techniques. *Journal of Applied Polymer Science*, 100, 1385-1394.

[121] Al-Ghezawi, N. Sanli,O. Isiklan, N. (2006) Permeation and separation characteristics of acetic acid-water mixtures by pervaporation through acrylonitrile and hydroxy ethyl methacrylate grafted poly(vinyl alcohol) membrane. *Separation Science and Technology*, 41, 2913-293.

[122] Kulkarni, S. S., Tambe, S. M., Kittur, A. A., Kariduraganavar, M. Y. (2006) Preparation of novel composite membranes for the pervaporation separation of water–acetic acid mixtures *Journal of Membrane Science*, 285, 420-431.

[123] Veerapur, R. S., Gudasi, K. B., Sairam, M., Shenoy, R. V., Netaji, M., Raju, K. V. S. N. Sreedhar, B., Aminabhavi, T. M. (2007) Novel sodium alginate composite membranes prepared by incorporating cobalt(III) complex particles used in pervaporation separation of water–acetic acid mixtures at different temperatures. *Journal of Material Science*, 42, 4406-4417.

[124] Namboodiri, V. V., Ponangi, R., Vane, L. M. (2006) A novel hydrophilic polymer membrane for the dehydration of organic solvents. *European Polymer Journal*, 42, 3390-3393.

[125] Ong, Y. T., Ahmad, A. L., Zein, S. H. S., Sudesh, K., Tan, S. H. (2011) Poly(3-hydroxybutyrate)-functionalized multi-walled carbon nanotubes/chitosan green nanocomposite membranes and their application in pervaporation, *Separation and Purification Technology*, 76, 419-427.

[126] Yeang, Q. W., Zein, S. H. S., Sulong, A. B., Tan, S. H. (2013) Comparison of the pervaporation performance of various types of carbon nanotube-based nanocomposites in the dehydration of acetone. *Separation Purification Technology*, 107, 252-263.

[127] Nakagawa, T. Kanemura, A. (1995) Synthesis and permeability of novel polymer membranes with high permselectivity for chlorinate hydrocarbons *Seni Gakaishi* 1995, 51, 123-130.

[128] Liu, Q. L., Xiao, H. (2004) Silicate-filled poly(siloxane imide) membranes for removal of VOCs from water by pervaporation. *Journal of Membrane Science*, 230, 121-129.

[129] Das, S., Banthia, A. K., Adhikari, B. (2006) Removal of chlorinated volatile organic contaminants from water by pervaporation using a novel polyurethane urea–poly (methyl methacrylate) interpenetrating network membrane. *Chemical Engineering Science*, 61, 6454-6467.

[130] Kim, K. S., Kwon, T. S., Yang, J. S., Yang, J. W. (2007) Simultaneous removal of chlorinated contaminants by pervaporation for the reuse of a surfactant. *Desalination*, 205, 87-96.

[131] Yu, J. Li, H. Liu, H. Z. (2006) Recovery of acetic acid over water by pervaporation with a combination of hydrophobic ionic liquids. *Chemical Engineering Communications*, 193, 1422-1430.

[132] Ghosh, U. K. Pradhan, N. C. Adhikari, B. (2006) Pervaporative recovery of *N*-methyl-2-pyrrolidone from dilute aqueous solution by using polyurethane urea membranes. *Journal of Membrane Science*, 285, 249-257.

[133] Ghosh, U. K. Pradhan, N. C. Adhikari, B. (2006) Synthesis and characterization of porous polyurethane urea membranes for pervaporative separation of 4-nitrophenol from aqueous solution. *Bulletin. Materials Science*, 291, 225-231.

[134] Das, S., Banthia, A. K., Adhikari, B. (2006) Pervaporation separation of DMF from water using a crosslinked polyurethane urea-PMMA IPN membrane. *Desalination*, 197, 106-116.

[135] Mujiburohman, M., Feng, X. S. (2007) Permselectivity, solubility and diffusivity of propyl propionate/water mixtures in poly (ether block amide) membranes. *Journal of Membrane Science*, 300, 95-103.

[136] Kujawa, J., Cerneaux, S., Kujawski, W. (2015) Removal of hazardous volatile organic compounds from water by vacuum pervaporation with hydrophobic ceramic membranes. *Journal of Membrane Science*, 474, 11-19.

[137] Kujawa, J., Cerneaux, S., Kujawski, W. (2015) Highly hydrophobic ceramic membranes applied to the removal of volatile organic compounds in pervaporation. *Chemical Engineering Journal*, 260, 43-54.

[138] Bai, Y. X., Qian, J. W., An, Q. F., Zhu, Z. H., Zhang, P. (2007) Pervaporation characteristics of ethylene–vinyl acetate copolymer membranes with different composition for recovery of ethyl acetate from aqueous solution. *Journal of Membrane Science,* 305, 152-159.

[139] Kujawski, W., Krajewski, S. R. (2007) Influence of inorganic salt on the effectiveness of liquid mixtures separation by pervaporation. *Separation and Purification Technology*, 57, 495-501.

[140] Mohammadi, T., Kikhavandi, T., Moghbeli, M. (2008) Synthesis and characterization of poly(ether-*block*-amide) membranes for the pervaporation of organic/aqueous mixtures, *Journal of Applied Polymer Science*, 107, 1917-1923.

[141] Uragami, T., Yamada, H., Miyata, T. (2001) Removal of dilute volatile organic compounds in water through graft copolymer membranes consisting of poly(alkyl methacrylate) and poly(dimethyl siloxane) by pervaporation and their membrane morphology. *Journal of Membrane Science,* 187, 255-269.

[142] Miyata, T., Yamada, H., T. Uragami, T. (2001) Surface modification of microphase-separated membranes by fluorine-containing polymer additive and removal of dilute benzene in water through these membranes, *Macromolecules,* 34, 8026-8033.

[143] Uragami, T., Yamada, H., T. Miyata, T. (2006) Effects of fluorine-containing graft- and block-copolymer additives on removal characteristics of dilute benzene in water by microphase-separated membranes modified with these additives, *Macromolecules*, 39,1890-1897.

[144] Uragami, T., Meotoiwa, T., Miyata, T. (2001) Effects of the addition of calixarene to microphase-separated membranes for the removal of volatile organic compounds from dilute aqueous solutions, *Macromolecules*, 34, 6806-6811.

[145] Uragami, T., Meotoiwa, T., Miyata, T. (2003) Effects of morphology of multicomponent polymer membranes containing calixarene on permselective removal of benzene from a dilute aqueous solution of benzene, *Macromolecules,* 36, 2041-2048.

[146] Han, X. Armstrong, D. W. (2007) Ionic liquids in separations, *Accounts of Chemical Research,* 40, 1079-1086.

[147] Uragami, T., Matsuoka, Y., Miyata, T. (2016) Permeation and separation characteristics in removal of dilute volatile organic compounds from aqueous solutions through copolymer membranes consisted of poly(styrene) and poly(dimethyl siloxane) containing hydrophobic ionic liquid by pervaporation. *Journal of Membrane Science*, 506, 109-118.

[148] Uragami, T., Fukuyama, E. Miyata, T. (2016) Selective removal of dilute benzene from water by poly (methyl methacrylate)-graft-poly (dimethyl siloxane) membranes containing hydrophobic ionic liquid by pervaporation. *Journal of Membrane Science*, 510, 131-140.

[149] Uragami, T., Ohshima, T., Miyata, T. (2003) Removal of Benzene from an Aqueous Solution of Benzene by Various Cross-linked Poly(dimethyl siloxane) Membranes during Pervaparation. *Macromolecules*, 36, 9430-9436.

[150] Uragami, T., Ohshima, T., Miyata, T. (2003) Removal of benzene from an aqueous solution of benzene by various cross-linked poly(dimethyl siloxane) membranes during pervaparation. *Macromolecules*, 36, 9430-9436.

[151] Ohshima, T., Miyata, T., Uragami, T. (2005) Selective removal of dilute benzene from water by various cross-linked poly(dimethyl siloxane) membranes containing *tert*-Butylcalix[4]arene, *Macromolecular Chemistry and Physics*, 206, 2521-2529.

[152] Juang, R. S., Lin, S. H., Yang, M. C. (2005) Mass transfer analysis on air stripping of VOCs from water in microporous hollow fibers. *Journal of Membrane Science*, 255, 79-87.

[153] Panek, D., Konieuny, K. (2007) Preparation and applying the membranes with carbon black to pervaporation of toluene from the diluted aqueous solutions. *Separation and Purification Technol*ogy, 57, 507-512.

[154] Ohshima, T., Kogami, Y., Miyata, T., Uragami, T. (2005) Pervaporation characteristics of cross-linked poly (dimethyl siloxane) membranes for removal of various volatile organic compounds from water. *Journal of Membrane Science*, 260, 156-163.

[155] Sun, L., Baker, G. L., Bruening, M. L. (2005) Polymer brush membranes for pervaporation of organic solvents from water. *Macromolecules*, 38, 2307-2314.

[156] Zhen, H. F., Jang, S. M. J., Teo, W. K., Li, K. (2006) Modified silicone–PVDF composite hollow-fiber membrane preparation and its application in VOC separation. *Journal of Applied Polymer Sci*ence, 99, 2497-2503.

[157] Panek, D., Konieczny, K. (2006) Pervaporation of toluene and toluene/acetone/ethyl acetate aqueous mixtures through dense composite polydimethylsiloxane membranes. *Desalination*, 200, 367-373.

[158] Panek, D., Konieczny, K. (2006) Pervaporation of toluene and toluene/acetone/ethyl acetate aqueous mixtures through dense composite polydimethylsiloxane membranes. *Desalination*, 200, 367-373.

[159] Inui, K., Okumura, H., Miyata, T., Uragami, T. (1997) Permeation and separation of benzene/cyclohexane mixtures through cross-linked poly(alkyl methacrylate) membranes *Journal of Membrane Science*, 132, 193-202.

[160] Inui, K. Okumura, H. Miyata, Uragami, T. (1997) Characteristics of permeation and separation of dimethyl acrylamide-methyl methacrylate random and graft copolymer membranes for a benzene/ cyclohexane mixture. *Polymer Bulletin*, 39, 733- 740.

[161] Inui, K., Miyata, T., Uragami, T. (1997) Permeation and separation of benzene/cyclohexane mixtures through liquid-crystalline polymer membranes, *Journal of Polymer Science, Part B: Polymer Phys*ics, 35, 699-707.

[162] Inui, K., Miyata, T., Uragami, T. (1998) Effect of permeation temperature on permeation and separation of a benzene/cyclohexane mixture through liquid-crystalline polymer membranes. *Journal of Polymer Science, Part B: Polymer Physics*, 36, 281-288.

[163] Inui, K., Miyata, T., Uragami, T. (1998) Permeation and separation of binary organic mixtures through a liquid-crystalline polymer membrane Permeation and separation of binary organic mixtures through a liquid-crystalline polymer membrane. *Macromolecular Chemistry and Physics*, 199, 589- 595.

[164] Inui K., Okazaki, K., Miyata, T., Uragami, T. (1998) Effect of mesogenic groups on characteristics of permeation and separation for benzene/cyclohexane mixtures of side-chain liquid-crystalline polymer membranes. *Journal of Membrane Science*, 143, 93-104.

[165] Uragami, T., Tsukamoto, K. Inui, T. Miyata, T. (1998) Pervaporation characteristics of a benzoylchiotsan membrane for benzene/ cyclohexane mixtures, *Macromolecular Chemistry and Physics*, 199, 49-54.

[166] Inui, K., Miyata, T., Uragami, T. (1998) Permeation and separation of a benzene/cyclohexane mixture through benzoylchitosan membranes, *Journal of. Membrane Science*, 138, 67-75.

[167] Uragami, T., Tsukamoto, K., Miyata, T., Heinze, T. (1999) Pervaporation characteristics for benzene/cyclohexane mixtures through benzoyl cellulose membranes, *Macromolecular Chemistry and Physics*, 200, 1985-1990.

[168] Uragami, T., Tsukamoto, K., Miyata, T., Heinze, T. (1999) Permeation and separation characteristics for benzene/cyclohexane mixtures through tosylcellulose membranes in pervaporation. *Cellulose*, 6, 221- 231.

[169] Uragami, T., Tsukamoto, K., Miyata, T., Heinze, T. (1999) Pervaporation characteristics for benzene/cyclohexane mixtures through benzoyl cellulose membranes. *Macromolecular Chemistry Physics*, 200, 1985-1990.

[170] Uragami, T., Tsukamoto, K., Miyata, T., Heinze, T. (1999) Permeation and separation characteristics for benzene/cyclohexane mixtures through tosylcellulose membranes in pervaporation. *Cellulose*, 6, 221- 231.

[171] Inui, K., Noguchi, T., Miyata, T., Uragami, T. (1999) Pervaporation characteristics of methyl methacrylate–methacrylic acid copolymer membranes ionically crosslinked with metal ions for a benzene/ cyclohexane mixture. *Journal of Applied Polymer Science*, 71, 233-241.

[172] Ren, J., Standt-Bickel, C., Lichtenthaler, R. (2001) Separation of aromatics/aliphatics with crosslinked 6FDA-based copolyimides *Separation and Purification Technol*ogy, 22-23, 31-43.

[173] Yildirim, A., Hilmiogle, N., Tulbentci, S. (2001) Pervaporation separation of benzene/cyclohexane mixtures by poly (vinyl chloride) membranes. *Chemical Engineering Technology*, 24, 275- 279.

[174] Pandey, L. K., Saxena, C., Dubey, V. (2003) Modification of poly (vinyl alcohol) membranes for pervaporative separation of benzene/cyclohexane mixtures. *Journal of Membrane Science*, 227, 173-182.

[175] Kusumocahyo, S. P., Ichikawa, T., Shinbo, T., Iwatsubo, T., Kameda, M., Ohi, K., Yoshimi, Y., Kanamori, T.(2005) Pervaporative separation of organic mixtures using dinitrophenyl group-containing cellulose acetate membrane. *Journal of Membrane Science*, 253. 43-48.

[176] Dubey, V., Pandey, L. K., Saxena, C. (2006) Pervaporation of benzene/cyclohexane mixtures through supramolecule containing poly (vinyl acetal) membranes. *Separation and Purification Technology*, 50, 45-50.

[177] Sun, H. L., Lu, L. Y., Peng, F. B., Wu, H., Jiang, Z. Y. (2006) Pervaporation of benzene/cyclohexane mixtures through CMS-filled poly (vinyl alcohol) membranes. *Separation and Purification Technology*, 52, 203 208.

[178] Lu, L. Y., Sun, H. L., Peng, F. B., Jiang, Z. Y. (2006) Novel graphite-filled PVA/CS hybrid membrane for pervaporation of benzene/ cyclohexane mixtures. *Journal of Membrane Science*, 281, 245-252.

[179] Peng, F. B. Jiang, Z. Y. Hu, C. L. Wang, Y. Q. Lu, L. Y. Wu, H. (2006) Pervaporation of benzene/cyclohexane mixtures through poly (vinyl alcohol) membranes with and without β-cyclodextrin. *Desalination*, 193, 182-192.

[180] Peng, F. B., Hu, C. L., Jiang, Z. Y. (2007) Novel poly (vinyl alcohol)/carbon nanotube hybrid membranes for pervaporation separation of benzene/cyclohexane mixtures. *Journal of Membrane Science*, 297, 236-242.

[181] Luo, Y. J., Xin, W., Li, G. P., Yang, Y., Liu, J. R., Lv, Y., Jiu, Y. B. (2007) Pervaporation properties of EC membrane crosslinked by hyperbranched-polyester acrylate. *Journal of Membrane Sci*ence, 303, 183-193.

[182] Zhang, X., Qian, L., Wang, H., Wei Zhong, Du, Q. (2008) Pervaporation of benzene/cyclohexane mixtures through rhodium-loaded b-zeolite-filled polyvinyl chloride hybrid membranes. *Separation and Purification Technology*, 63, 434-443.

[183] Shen A, J. N., Chu, Y. X., Ruan, H. M., A, Wu, L. G., Gao, C. J., Van der Bruggen, B. (2014) Pervaporation of benzene/cyclohexane mix-

tures through mixed matrix membranes of chitosan and Ag^+/carbon nanotubes. *Journal of Membrane Science*, 462,15 160-169.

[184] Okuno, H., Nishida, Uragami, T. (1995) Preferencial absorption and permeation of binary mixtures in poly(vinyl chloride) membrane by pervaporation. *Journal of Polymer Science, Part B: Polymer Phys*ics, 33, 299-307.

[185] Mulder, M. H. V. Franken, T., Smolders, C. A. (1985) Preferential absorption versus preferential permeability in pervaporation. *Journal of Membrane Scienc*e, 22, 155-173.

[186] Ruckenstein, E., Sun, F. (1995) Hydrophobic-hydrophilic composite membranes for the pervaporation of benzene-ethanol mixtures. *Journal of Membrane Sci*ence, 103, 271-283.

[187] Zhou, M., Persin, M., Kujawski, W. J., Sarrazin, J. (1995) Electrochemical preparation of polypyrrole membranes and their application in ethanol-cyclohexane separation by pervaporation. *Journal of Membrane Sci*ence, 108, 89-96.

[188] Johnson, T., Thomas, S. (1999) Pervaporation of acetone-chlorinated hydrocarbon mixtures through polymer blend membranes of natural rubber and epoxidized natural rubber. *Journal of Applied Polymer Science*, 71, 2365- 2379.

[189] Gorri, D., Ibanez, R., Ortiz, I. (2006) Comparative study of the separation of methanol–methyl acetate mixtures by pervaporation and vapor permeation using a commercial membrane. *Journal of Membrane Science*, 280, 582-593.

[190] Matsui, S., Paul, D. R. (2002) Pervaporation separation of aromatic/aliphatic hydrocarbons by crosslinked poly(methyl acrylate-*co*-acrylic acid) membranes. *Journal of Membrane Science*, 195, 229-245.

[191] Aouinti, L., Roizard, D., Hu, G. H., Thomas, F., Belbachir, M. (2009) Investigation of pervaporation hybrid polyvinylchloride membranes for the separation of toluene–*n*-heptane mixtures — case of clays as filler. *Desalination*, 241, 174-181.

[192] Wu, T., Wang, N., Li, J., Wang, L., Zhang, W., Zhang, G., Ji, S. (2015) Tubular thermal crosslinked-PEBA/ceramic membrane for aromatic/aliphatic pervaporation. *Journal of Membrane Science*, 486, 1-9].

[193] Zhang, Y., Wang, N., Ji, S., Zhang, R., Zhao, C., Li, J. R. (2015) Metal–organic framework/poly(vinyl alcohol) nanohybrid membrane for the pervaporation of toluene/*n*-heptane mixtures. *Journal of Membrane Science*, 489, 144-152.

[194] Zhang, Y., Wang, N., Zhao, C., Wang, L., Ji, S., Li, J. R. (2016) $Co(HCOO)_2$-based hybrid membranes for the pervaporation separation of aromatic/aliphatic hydrocarbon mixtures. *Journal of Membrane Science*, 520, 646-656.

[195] Wang, L., Han, X., Li, J., Zhan, X., Chen, J. (2011) Hydrophobic nano silica/polydimethylsiloxane membrane for dimethyl carbonate–methanol separation via pervaporation. *Chemical Engineering Journal*, 171, 1035-1044.

[196] Kononova, S. V., Kremnev, R. V., Suvorova, E. I., Baklagina, Y. G., Volchek, B. Z., Uchytil, P., Shabsels, B. M., Romashkova, K. A., Setnickova, K., Reznickova, J. (2015) Pervaporation membranes with poly(γ-benzyl-L-glutamate) selective layers for separation of toluene–*n*-heptane mixtures. *Journal of Membrane Science*, 477, 14-24.

[197] Wang, N., Wu, T., Wang, L., Li, X., Zhao, C., Li, J., Ji, S. (2017) Hyperbranched polymer composite membrane using water as solvent for separating aromatic/aliphatic hydrocarbon mixtures. *Separation and Purification Technology*, 179, 225-235.

[198] Ray, S., Ray, S. K. (2006) Separation of organic mixtures by pervaporation using crosslinked and filled rubber membranes *Journal of Membrane Science*, 285, 108-119.

[199] Chakraborty, M., Bart, H. J. (2007) Highly selective and efficient transport of toluene in bulk ionic liquid membranes containing Ag + as carrier. *Fuel Processing Technology*, 88, 43-49.

[200] Hasanoglu, A., Salt, Y., Keleser, S., Ozkan, S., Dincer, S. (2007) *Chemical Engineering and Processing: Process Intensification*, 46, 300-306.

[201] Wang, L. Y., Li, J. D., Lin, Y. Z., Chen, C. X. (2007) Separation of dimethyl carbonate/methanol mixtures by pervaporation with poly (acrylic acid)/poly (vinyl alcohol) blend membranes. *Journal of Membrane Science*, 305, 238-246.

[202] Wang, Y., Hirakawa, S., Tanaka, K., Kita, H., Okamoto, K. (2003) Pervaporation properties of aromatic/nonaromatic hydrocarbons of crosslinked membranes of copolymers based on diethyl vinylbenzyl-phosphonate. *Journal of Applied Polymer Science,* 87, 2177-2185.

[203] Schwarz, H. H., Malsch, G. (2005) Polyelectrolyte membranes for aromatic–aliphatic hydrocarbon separation by pervaporation. *Journal of Membrane Science*, 247, 143-152.

[204] Shen, J. N., Zheng, X. C., Ruan, H. M., Wu, L. G., Qiu, J. H., Gao, C. J. (2007) Synthesis of AgCl/PMMA hybrid membranes and their absorption performance of cyclohexane/cyclohexene. *Journal of Membrane Science*, 304, 118-124.

[205] Branco, L. C., Crespo, J. G., Afonso, C. A. M.(2002) Studies on the selective transport of organic compounds by using ionic liquids as novel supported liquid membranes. *Chemistry - A. European Journal,* 8, 3865-3871.

[206] Berendsen, W., Radmer, P., Reuss, M. (2006) Pervaporative separation of ethanol from an alcohol—ester quaternary mixture. *Journal of Membrane Science*, 280, 684-692.

[207] Matsumoto, M. Inomoto, Y. Kondo, K.(2005) Selective separation of aromatic hydrocarbons through supported liquid membranes based on ionic. *Journal of Membrane Science*, 246, 77-81.

[208] Qi, R. B. Wang, Y. J. Li, J. D. Zhao, C. W. Zhu, S. L. (2006) Pervaporation separation of alkane/thiophene mixtures with PDMS membrane. *Journal of Membrane Science*, 280, 545-552.

[209] Park, H. C. Ramaker, N. E. Mulder, M. H. V. Smolders, C. A. (1995) Separation of MTBE-methanol mixtures by pervaporation. *Separation Science and Technol*ogy, 30, 419-433.

[210] Chen, W. J., Martin, C. R. (1995) Highly methanol-selective membranes for the pervaporation separation of methyl *t*-butyl

ether/methanol mixtures. *Journal of Membrane Science*, 104, 101-108.

[211] Jonquieres, A., Roizard, D., Lochon, P. (1995) Contribution of absorption to global mass transfer during pervaporation of ethyl *tert*-butyl ether–ethanol mixtures through a polyurethaneimide film. *Journal of the Chemical Society, Faraday Transaction,* 91, 1247-1251.

[212] Cai, B. X., Yu, L. Ye, H. Gao, C. J. (2001) Effect of separating layer in pervaporation composite membrane for MTBE/MeOH separation. *Journl of Membrane Science*, 194, 151-156.

[213] Schwarz, H. H., Apostel, R., Paul, D. (2001) Membranes based on polyelectrolyte–surfactant complexes for methanol separation. *Journal of Membrane Science*, 194, 91-102.

[214] Niang, M., Luo, G. S. (2001) A triacetate cellulose membrane for the separation of methyl *tert*-butyl ether/methanol mixtures by pervaporation. *Separation and Purification Technology*, 24, 427-435.

[215] Hilmioglu, N. D., Tulbentci, S. (2004) Pervaporation of MTBE/ methanol mixtures through PVA membranes. *Desalination,* 160, 263-270.

[216] Ray, S., Ray, S. K. (2006) Synthesis of highly methanol selective membranes for separation of methyl tertiary butyl ether (MTBE)–methanol mixtures by pervaporation. *Journal of Membrane Science*, 278, 279-289.

[217] Touchal, S., Roizard, D., Perrin, L. (2006) Pervaporation properties of polypyrrolidinone-based membranes for EtOH/ETBE mixtures separation. *Journal of Applied Polymer Science*, 99, 3622-3630.

[218] Wang, Y., Chung, T. S., Wang, H., Goh, S. H. (2009) Butanol isomer separation using polyamide–imide/CD mixed matrix membranes via pervaporation. *Chemical Engineering Science*, 64, 5198-5209.

[219] Ishihara, K., Matsui, K., Fujii, H., Nishide, H., Shinohara, I. (1985) Separation of xylene isomer by pervaporation through a highly permselective polymer membrane having dinitrophenyl group. *Chemistry Letters*, 11, 1663-1666.

[220] Miyata, T., Iwamoto, T., Uragami, T. (1994) Characteristics of permeation and separation for propanol iosmers through poly(vinyl alcohol) membranes containing cyclohexane. *Journal of Applied Polymer Science,* 51, 2007-2014.

[221] Miyata, T., Iwamoto, T., Uragami, T. (1996) Characteristics of permeation and separation of xylene isomers through poly (vinyl alcohol) membranes containing cyclodextrin. *Macromolecular Chemistry Physics*, 197, 2909-2921.

[222] Kusumocahyo, S. P. Kanamori, T. Sumaru, K. Iwatsubo, T. Shinbo, T. (2004) Pervaporation of xylene isomer mixture through cyclodextrins containing polyacrylic acid membranes. *Journal of Membrane Science*, 231, 127-132.

[223] Touil, S., Tingry, S., Palmeri, J., Bouchtalla, S., Deratani, A.(2005) Preparation and characterization of α-cyclodextrin-containing membranes—application to the selective extraction of xylene isomers. *Polymer*, 46, 9615-9625.

[224] Kusumocahyo, S. P. Ichikawa, T., Shinbo, T., Iwatsubo, T., Kameda, M., Ohi, K., Yoshimi, Y., Kanamori, T. (2005) Pervaporative separation of organic mixtures using dinitrophenyl group-containing cellulose acetate membrane. *Journal of Membrane Science*, 253. 43-48.

[225] Touil, S.,Tingry, S., Bouchtalla, S., Deratani, A. (2006) Selective pertraction of isomers using membranes having fixed cyclodextrin as molecular recognition sites. *Desalination*, 193, 291-298.

[226] Lue, S. J., Liaw, T. H. (2006) Separation of xylene mixtures using polyurethane-zeolite composite membranes. *Desalination*, 193, 137-143.

[227] Zhang, P., Qian, J. W., Yang, Y., Bai, Y. X., An, Q F., Yan, W. D. (2007) Swelling behavior of palygorskite–polyacrylamide hybrid membrane in xylene mixtures and its pervaporation performance for separating the xylene isomers. *Journal of Membrane Science*, 288, 280-289.

[228] Wang, Y., Chung, T. S., Wang, H., Goh, S. H. (2009) Butanol isomer separation using polyamide–imide/CD mixed matrix membranes via pervaporation. *Chemical Engineering Science*, 64, 5198-5206.

[229] Chen, H. L., Wu, L. G., Tan, J., Zhu, C. L. (2000) PVA membrane filled β-cyclodextrin for separation of isomeric xylenes by pervaporation. *Chemical Engineering Journal*, 78,159-164.

[230] Qu, X. Y., Dong, H., Zhou, Z. J., Zhang, L., Chen, H. L. (2010) Pervaporation separation of xylene isomers by hybrid membranes of PAAS filled with silane-modified zeolite. *Industrial Engineering Chemistry Research*, 49, 7504–7514.

[231] Zheng, H., Yoshikawa, M. (2015) Molecularly imprinted cellulose membranes for pervaporation separation of xylene isomers. *Journal of Membrane Science,* 478, 148-154.

[232] Editors-in-Chief: Drioli, E., Giorno, L., Fontananova, E. (2017) *Comprehensive Membrane Science and Engineering* (Second Edition) Elsevier].

[233] Feng, X., Huang, R. Y. M. (1996) Studies of a membrane reactor: Esterification facilitated by pervaporation. *Chemical Engineering Science*, 51, 4673-4679.

[234] Genduso, G., Luis, P., Van der Bruggen, B. (2015) 19–Pervaporation membrane reactors (PVMRs) for esterification. *Membrane Reactors for Energy Applications and Basic Chemical Production.* A volume in Woodhead Publishing Series in Energy, Pages 565–603.

[235] Chandane, V. S., Rathod, A. P., Kailas L. Wasewar, K. L. (2016) Enhancement of esterification conversion using pervaporation membrane reactor. *Resource-Efficient Technologies*, 2, S47-S52.

[236] Castanheiro, J. E., Ramos, A. M., Fonseca, I. M., Vitai, J. T. (2006) Esterification of acetic acid by isoamylic alcohol over catalytic membranes of poly(vinyl alcohol) containing sulfonic acid groups. *Applied Catalysis A: General*, 311, 17-23.

[237] Benedict, D. J., Parulekar, S. J., Tsai, S. P. (2006) Pervaporation-assisted esterification of lactic and succinic acids with downstream ester recovery. *Journal of Membrane Science,* 281, 435-445.

[238] Qing, W., Chen, J., Shi, X., Wu, J., Hu, J., Zhang, W. (2017) Conversion enhancement for acetalization using a catalytically active membrane in a pervaporation membrane reactor. *Chemical Engineering Journal*, 313, 1396-1405.

[239] Sanz, M. T., Gmehling, J. (2006) Esterification of acetic acid with isopropanol coupled with pervaporation: Part II. Study of a pervaporation reactor. *Chemical Engineering Journal*, 123, 9-14.

[240] Matsui, S., Paul, D. R. (2002) Pervaporation separation of aromatic/aliphatic hydrocarbons by crosslinked poly(methyl acrylate-*co*-acrylic acid) membranes. *Journal of Membrane Science*, 195, 229-245.

[241] Aouinti, L., Roizard, D., Hu, G. H., Thomas, F., Belbachir, M. (2009) Investigation of pervaporation hybrid polyvinylchloride membranes for the separation of toluene–*n*-heptane mixtures — case of clays as filler. *Desalination*, 241, 174-181.

[242] Rathod, A. P., Wasewar, K. L., Sonawane, S. S. (2013) Enhancement of esterification reaction by pervaporation reactor: An intensifying approach. *Procedia Engineering*, 51, 330-334.

[243] Wu, T., Wang, N., Li, J., Wang, L., Zhang, W., Zhang, G., Ji, S. (2015) Tubular thermal crosslinked-PEBA/ceramic membrane for aromatic/aliphatic pervaporation. *Journal of Membrane Science*, 486, 1-9.

[244] Zhang, Y., Wang, N., Ji, S., Zhang, R., Zhao, C., Li, J. R. (2015) Metal–organic framework/poly(vinyl alcohol) nanohybrid membrane for the pervaporation of toluene/*n*-heptane mixtures. *Journal of Membrane Science*, 489, 144-152.

[245] Zhang, Y., Wang, N., Ji, S., Zhang, R., Zhao, C., Li, J. R. (2015) Metal–organic framework/poly(vinyl alcohol) nanohybrid membrane for the pervaporation of toluene/*n*-heptane mixtures. *Journal of Membrane Science*, 489, 144-152.

[246] Wang, L., Han, X., Li, J., Zhan, X., Chen, J. (2011) Hydrophobic nano silica/polydimethylsiloxane membrane for dimethyl carbonate–methanol separation via pervaporation. *Chemical Engineering Journal*, 171, 1035-1044].

[247] Zhang, Y., Wang, N., Ji, S., Zhang, R., Zhao, C., Li, J. R. (2015) Metal–organic framework/poly(vinyl alcohol) nanohybrid membrane for the pervaporation of toluene/*n*-heptane mixtures. *Journal of Membrane Science*, 489, 144-152.

[248] Wang, N., Wu, T., Wang, L., Li, X., Zhao, C., Li, J., Ji, S. (2017) Hyperbranched polymer composite membrane using water as solvent for separating aromatic/aliphatic hydrocarbon mixtures. *Separation and Purification Technology*, 179, 225-235.

[249] Chen, M. S., Eng, R. M., Glazer, J. L., Charles G. Wensley, C. G. (1988) *Pervaporation process for separating alcohols from ethers.* US4774365A.

[250] Shi-kuan, M., Michael, R., Leonard, J., Glen, C. (1990) *Pervaporation process for separating alcohols from ethers.* European Patent Application EP0317918.

[251] Aiouache, F., Goto, S. (2003) Reactive distillation–pervaporation hybrid column for *tert*-amyl alcohol etherification with ethanol. *Chem-ical Engineering Science*, 58, 2465-2477.

[252] Yang, B. L., Goto, S. (2006) Pervaporation with reactive distillation for the production of ethyl *tert*-butyl ether. *Separation Science and Technology*, 32, 971-981.

[253] Yee, K. F., Ong, Y. T., Mohamed, A. R., Tan, S. H. (2014) Novel MWCNT-buckypaper/polyvinyl alcohol asymmetric membrane for dehydration of etherification reaction mixture: fabrication, characterization and application. *Journal of Membrane Science*, 453, 546-555.

[254] Frusteri, F., Cannilla, C., Bonura, G., Frusteri. L. (2014) Glycerol etherification with TBA: high yield to poly-ethers using a membrane assisted batch reactor. *Environmental Science & Technology*, 48, 6019-026.

[255] Cannilla· C., Bonura· G., Frusteri· L., Frusteri· F. (2015) Batch reactor coupled with water permselective membrane: Study of glycerol etherification reaction with butanol. *The Chemical Engineering Journal*, 282, 187-193.

[256] Ca Shi-kuan, M. Michael, R., Leonard, J., Glen, C. (1993) *Pervaporation process for separating alcohols from ethers*. European Patent EP0317918B1.

[257] Khudsange, C. R., Wasewar, K. L. (2017) Process intensification of esterification reaction for the production of propyl butyrate by pervaporation. *Resource-Efficient Technologies*, 3, 88-93.

In: Pervaporation
Editor: Jean Garcia
ISBN: 978-1-53614-459-8

Chapter 3

HIGH PERFORMANCE BIO-BASED MEMBRANES FOR BIOFUEL PURIFICATION BY PERVAPORATION

Anne Jonquieres*[*]*, Carole Arnal-Herault, Jérôme Babin, Magali Billy and Faten Hassan Hassan Abdellatif
Laboratoire de Chimie Physique Macromoléculaire,
Université de Lorraine/CNRS, Nancy, France

ABSTRACT

Ethyl *tert*-butyl ether (ETBE) is a major biofuel mainly produced from the reaction of isobutene with a large excess of bioethanol. This ether greatly improves fuel combustion and limits the emission of toxic hydrocarbons. Moreover, thanks to its particular molecular structure, this ether does not accumulate into the environment contrary to methyl *tert*-butyl ether (MTBE). Nevertheless, during the industrial ETBE synthesis, this ether forms an azeotropic mixture containing 20 wt% of ethanol, which must be removed during the final purification stage. The ternary distillation process generally used for this separation is highly energy intensive.

[*] Corresponding Author Email: anne.jonquieres@univ-lorraine.fr.

Several works have shown that the pervaporation membrane process (PV) could be a very good alternative for ETBE purification with very important energy savings. This chapter focuses on our on-going development of high performance bio-based cellulosic membranes for ETBE purification by PV. Cellulose acetate (CA) is extremely selective for ethanol removal from ETBE but its flux is very low. Different strategies for greatly improving its flux while maintaining a very high selectivity are described and the main relationships between membrane structure, morphology and properties are illustrated. In a first part, the grafting of CA by controlled radical polymerization enables to vary the length and the number of polymethacrylate grafts, inducing different copolymer morphologies and PV properties. In particular, it is shown that, for a same graft content in the bio-based membranes, the nanostructuration of the grafts plays a determining role on the separation features. The second part explores the development of purely bio-based membranes made of a CA backbone with bio-based polylactide grafts based on a controlled grafting by "click" chemistry. This strategy enables to plasticize the cellulosic membranes very efficiently and greatly improves the membrane properties. In the third part, the grafting of ionic liquids onto CA is reported as another very efficient way of inducing CA plasticization. The influence of the ionic liquid content and molecular structure is discussed to reveal the key parameters for the membrane properties. Several bio-based membranes with grafted ionic liquids have very improved pervaporation flux (up to almost 20 times that of virgin CA) and permeate ethanol only, corresponding to outstanding infinite separation factors for this challenging liquid/liquid separation.

1. INTRODUCTION

Ethyl *tert*-butyl ether (ETBE) is a major biofuel mainly produced from the reaction of isobutene with a large excess of bio-ethanol. This bulky ether is extensively used as an oxygenated additive to greatly improve fuel combustion and limits the environmental emission of toxic hydrocarbons [1, 2] (Figure 1). ETBE has also played a great role in the replacement of the former toxic lead derivatives. Moreover, thanks to its particular molecular structure, this ether does not accumulate into the environment contrary to methyl *tert*-butyl ether (MTBE). Nevertheless, during the industrial ETBE synthesis, this ether forms an azeotropic mixture containing 20 wt% of ethanol, which must be removed during the final purification stage. The ternary distillation process generally used for this separation is highly energy intensive [2-4]. Several works have shown that the pervaporation membrane process (PV) or distillation/PV hybrid processes could be very good alternatives for ETBE purification with very important energy savings compared to ternary distillation [5-9].

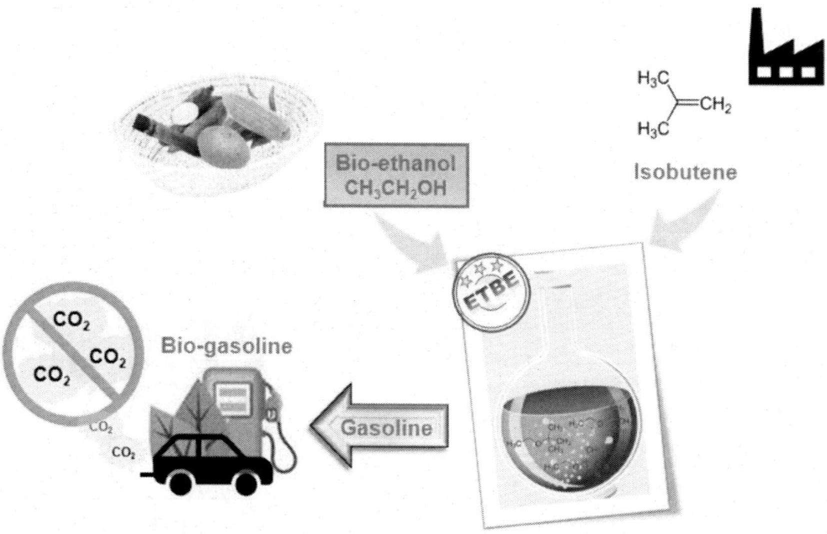

Figure 1. Importance of the ethyl *tert*-butyl biofuel.

The development of PV membranes for ETBE purification is quite challenging and the target membranes should be able to extract ethanol from the azeotrope ETBE/ethanol with high selectivity and high flux [2, 4, 8, 10]. PV membranes for ETBE purification remain relatively rare in the literature and their properties have been reviewed in the past few years [3, 11-14]. Three main types of polymeric membranes have been reported for this application. The first ones by order of importance are bio-based cellulosic membranes. Polymer blends based on poly(vinyl pyrrolidone) (PVP) and segmented multi-block copolymers have also led to interesting membrane properties for ETBE purification by PV.

Table 1. Cellulosic esters, related cellulosic polymer blends and semi-interpenetrated networks for the separation of the azeotropic mixture ETBE/EtOH (80/20 wt/wt %) by pervaporation (adapted from Table 1 of [11] with permission from Elsevier)

Membrane	T (°C)	J (kg/m² h)	C'_{EtOH} (wt%)	Ref.
Cellulosic esters :				
CA	40	0.08	≈ 100	[15]
CAP	40	0.70	96	[15]
CAB	40	1.7	90	[15]
Cellulosic blends :				
CAB-CAP (10-90 wt/wt %)	40	0.6	96.7	[18]
CAB-CAP (20-80 wt/wt %)	40	0.7	95.7	[18]
CAB-CAP (40-6 wt/wt %)	40	1.3	93.3	[18]
CAB-CAP (50-50 wt/wt %)	40	3.0	90.7	[18]
CAP-PAA (95-5 wt/wt %)	40	0.82	96.4	[20]
CAP-PAA (80-20 wt/wt %)	40	0.87	95.2	[20]
CAP-PAA (70-30 wt/wt %)	40	0.88	95.0	[20]
CA-P(VP-co-VAc) (40-60 wt/wt %)	40	2.3	86.0	[19]
CA- P(VP-co-VAc) (60-40 wt/wt %)	40	0.8	97.5	[19]
CA- P(VP-co-VAc) (95-5 wt/wt %)	40	0.3	98.0	[19]
CAP-P(VP-co-AA) (95-5 wt/wt %)	40	0.52	97.7	[21]
CAP-P(VP-co-AA) (95-5 wt/wt %)	50	0.87	96.6	[21]
CAP-P(VP-co-AA) (95-5 wt/wt %)	60	1.24	95.9	[21]
Cellulosic semi-IPNs				
CA-EGDMA (50-50 mol/mol %)	40	0.02	1	[15]
CAP-EGDMA (50-50 mol/mol %)	40	2.0	0.96	[15]
CAB-EGDMA (50-50 mol/mol %)	40	4.4	0.895	[15]
CA-PEG200DMA (50-50 mol/mol %)	40	0.02	1	[15]
CA-TEEMA.PEG200DMA (50-25.25 mol/mol %)	40	0.50	0.97	[15]
CA-TEEMA.PEG600DMA (50-25.25 mol/mol %)	40	0.90	0.96	[15]

J: normalized flux for a reference thickness of 5 microns. C'_{EtOH}: ethanol content (wt%) in the permeate. CA: cellulose acetate, CAP: cellulose acetate propionate, CAB: cellulose acetate butyrate, PAA: poly(acrylic acid), P(VP-co-Vac): poly(vinyl pyrrolidone-co-(vinyl acetate)), P(VP-co-AA): poly(vinyl pyrrolidone-co-(acrylic acid)), EGDMA: ethyleneglycol dimethacrylate, PEG200DMA: polyethyleneglycol 200 dimethacrylate, PEG600DMA: polyethyleneglycol 600 dimethacrylate, TEEMA: triethoxy ethyl methacrylate.

Bio-based cellulosic membranes appear particularly promising for the target application because their selectivity is generally high and their flux can be varied to a great extent by playing with the macromolecular structure. Cellulosic esters have shown quite outstanding properties for ETBE purification [15-17] (Table 1). In particular, cellulose acetate (CA) was extremely selective with an ethanol permeate content C'_{EtOH} of 100 wt% but it flux of 0.08 kg/m^2 h for a reference thickness of 5 microns was fairly low at 40 °C. Cellulose acetate propionate (CAP) and butyrate (CAB) had higher fluxes but lower selectivity, which still remained in the high range for ETBE purification in these conditions ($C'_{EtOH} \geq 90$ wt%) [15-17]. Cellulosic ester blends enabled to vary the fluxes over a broader range (0.6 - 3 kg/m^2 h) while still maintaining high ethanol permeate contents (96.7 - 90.7 wt%) [18]. Other cellulosic polymer blends have also been reported with interesting fluxes but the membrane selectivity was significantly decreased compared to that of cellulose acetate [18-21] (Table 1). Nevertheless, it was sometimes reported that some of the polymers involved in these cellulosic polymer blends were dissolved in the azeotropic mixture ETBE/ethanol during PV, leading to the decrease in the corresponding membrane properties with time [22]. A first solution to this problem was proposed with the development of semi-interpenetrated cellulosic/synthetic polymer networks (semi-IPNs). These semi-IPNs were made from cellulose acetate and cross-linked poly-(meth)acrylate networks obtained by photopolymerization. The corresponding membranes had stable properties but they were strongly anisotropic [15, 23].

In this chapter, we focus on our on-going development of new bio-based cellulosic membranes with high performance for ETBE purification by PV (Figure 2). As mentioned, cellulose acetate (CA) is extremely selective for ethanol removal from ETBE but its flux is too low for the industrial separation. Different strategies for greatly improving its flux while maintaining a very high selectivity will be described and the main relationships between the membrane structure, morphology and properties will be illustrated.

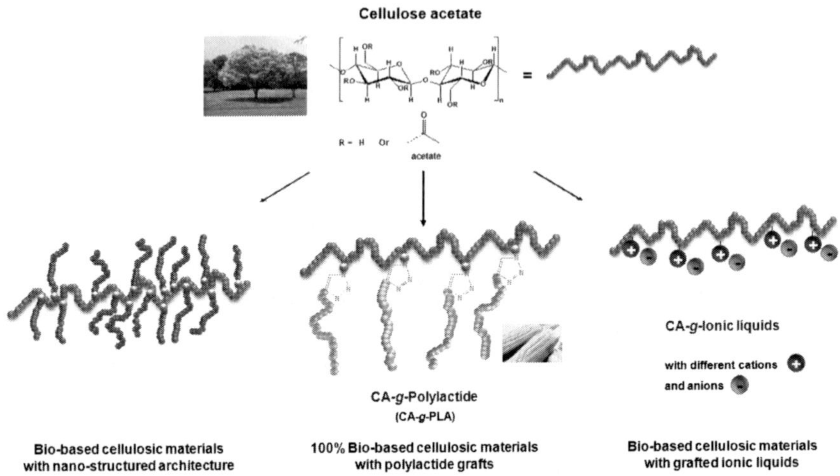

Figure 2. The different original bio-based cellulosic materials developed for ETBE purification by PV.

In the second part of this chapter, the grafting of CA by controlled radical polymerization will be explored to vary the length and the number of polymethacrylate grafts, inducing different copolymer morphologies and PV properties [11]. In particular, it will be shown that, for a same graft content in the bio-based membranes, the nanostructuration of the grafts plays a determining role on the separation features. The third part will describe the development of purely bio-based membranes made of a CA backbone and bio-based polylactide grafts, based on a controlled grafting by "click" chemistry [13]. This strategy enabled to plasticize the cellulosic membranes very efficiently and greatly improved the membrane properties. In the fourth part, the grafting of ionic liquids onto CA will be reported as another very efficient way of inducing CA plasticization [14, 24]. The influence of the ionic liquid content and molecular structure will be discussed to reveal the key parameters for the membrane properties. The last strategy led to several bio-based cellulosic membranes capable of permeating ethanol only, with very improved flux (up to almost 20 times that of virgin CA) and outstanding infinite separation factors for ETBE purification by PV.

2. Bio-Based Cellulosic Materials with Nano-Structured Architecture for ETBE Purification by Pervaporation

2.1. Strategy for the Synthesis of the New Bio-Based Membrane Materials

In the first strategy for developing new bio-based membranes for ETBE purification by PV, cellulose acetate was grafted with poly(methyl diethylene glycol methacrylate) (PMDEGMA) by controlled radical polymerization (Atom Transfer Radical Polymerization - ATRP) [25] (Figure 3). PMDEGMA grafts contain ethylene oxide hydrophilic units ($-CH_2CH_2O-$), which are known for their strong affinity with ethanol according to the former works reported in Table 1. The cellulose acetate used in our work was made from glycosidic rings containing 2.47 acetyl groups and 0.53 hydroxyl groups. Some of these hydroxyl groups were chemically modified to introduce brominated initiator groups for the ATRP. The polymer grafts were then obtained by polymerizing the methacrylate monomer from these initiator groups by ATRP.

This “grafting from” method enabled to precisely control the number and the length of the polymethacrylate grafts. Two series of grafted cellulosic materials with different architectures were thus obtained with either long grafts or small grafts and almost the same graft weight contents. The corresponding abbreviations were GLX and GSX for the copolymers containing X wt% of long grafts and short grafts, respectively.

A synchrotron WAXS and SAXS study of the grafted cellulosic materials showed that they were not crystalline (which is an advantage for membrane flux) and that their architecture had a strong influence on their morphology and the dispersion state of their grafts (Figure 4). The short grafts were homogeneously dispersed in the cellulosic materials and, consequently, they were highly constrained by the rigid cellulosic chains. On the contrary, the long grafts tended to induce a phase separation at the nano-scale, which increased with the graft content. Therefore, the grafted copolymer with the highest content of long grafts (GL44) displayed a fully segregated morphology, which corresponded to graft nano-domains with an average radius of gyration of 11.5 nm. In this particular copolymer (GL44), these highly permeable nano-domains were forming a soft percolating network with important consequences on the membrane properties for ETBE purification by PV.

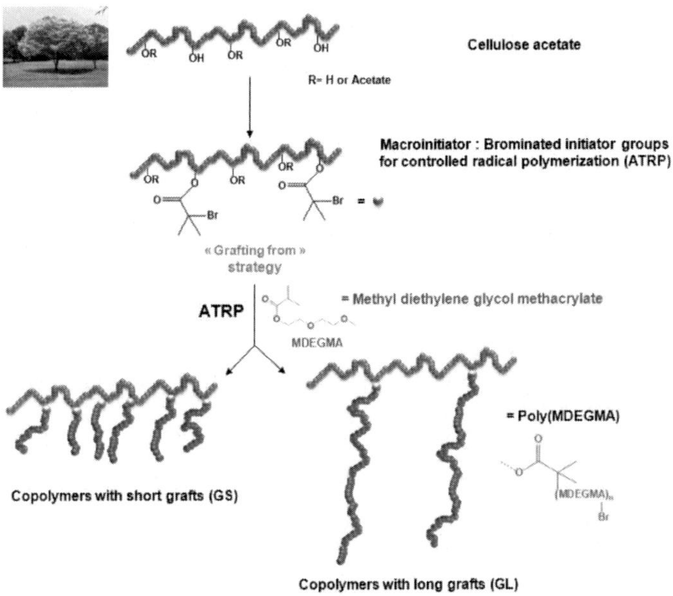

Figure 3. Strategy for the synthesis of new bio-based cellulosic materials with nano-structured architecture for ETBE purification by pervaporation.

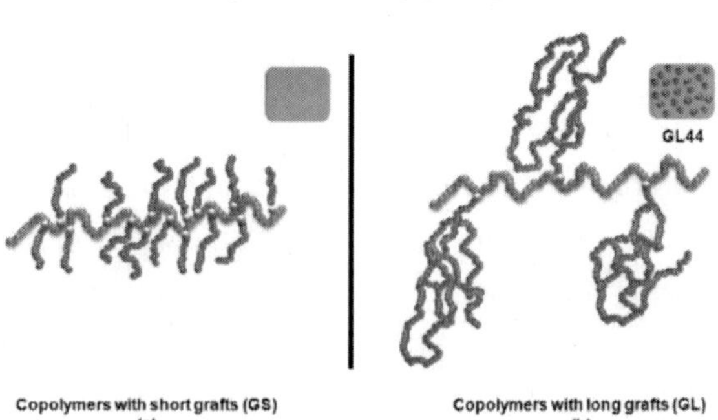

Figure 4. Influence of the grafted copolymers architecture on the membrane material morphology and the nano-structuration of the grafts (adapted from Figure 3 of [11] with permission from Elsevier).

2.2. Structure-Membrane Property Relationships for ETBE Purification by PV

The membrane properties were assessed at 50 °C for the PV separation of the target azeotropic mixture ETBE/ethanol (80/20 wt%/wt%) for the two series of grafted copolymer membranes (GLX and GSX) and virgin CA (i.e., the non-grafted cellulosic material) for comparison. For virgin CA, a low PV flux of 0.018 kg/m^2 h for a reference thickness of 5 microns was obtained as expected and the ethanol weight percent in the permeate was 98.8. The grafting of CA induced a strong improvement of the membrane properties. Furthermore, it turned out that, for a same graft content, the membrane properties strongly differed for the copolymers with either long or short grafts (Figures 5 and 6). These results first revealed the importance of the grafted copolymer architecture on the membrane flux and selectivity.

The PV flux of the grafted copolymers with short grafts (GSX) increased linearly with their graft content (Figure 5). This linear increase was in good agreement with the materials homogeneity at the nano-scale, which resulted from the good dispersion of their short grafts. Nevertheless, Figure 5 shows that the flux increase was limited by the strong constraints of the rigid polysaccharide chains on the short grafts. On the contrary, the PV flux of the grafted copolymers with long grafts (GLX) were systematically higher than those obtained with the short grafts and it also increased much more strongly with the graft content. In these copolymers GLX, the nano-structuration of the long grafts was responsible for their higher mobility, which induced better membrane permeability. This was particularly obvious for the copolymer with the highest content of long grafts (GL44). In the latter case, the polymer grafts were fully segregated and formed a soft percolating network responsible for a very strong increase in permeability.

For both series of grafted copolymers, the ethanol content in the permeate decreased when the PV flux increased as expected from the permeability/selectivity trade-off (Figure 6). Nevertheless, this decrease was very small for the copolymers with the short grafts (GSX). For these copolymers, the strong constraints of the rigid polysaccharide chains on the short grafts strongly limited the mass transfer of the bulky ETBE molecules.

Consequently, all of the membranes containing short grafts were highly selective for ethanol removal from the azeotropic mixture ETBE/ethanol by PV.

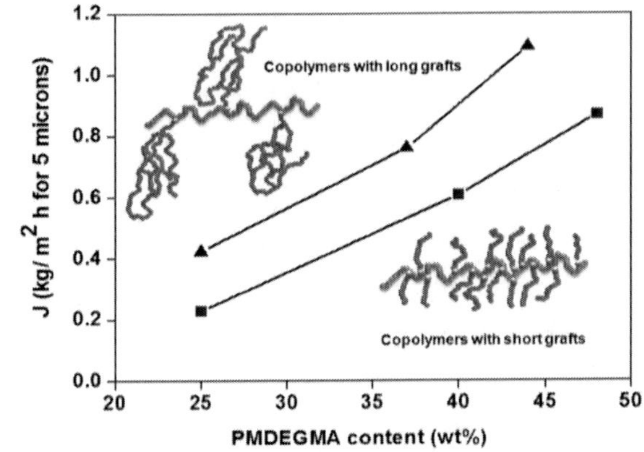

Figure 5. Influence of the copolymer graft content and architecture on the membrane flux for the PV separation of the azeotropic mixture ETBE/EtOH at 50 °C (reproduced from Figure 6 of [11] with permission from Elsevier).

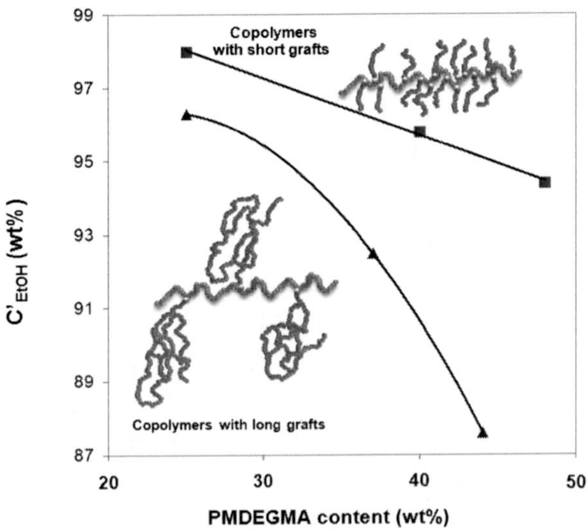

Figure 6. Influence of the copolymer graft content and architecture on the ethanol weight fraction in the permeate for the PV separation of the azeotropic mixture ETBE/EtOH at 50 °C (reproduced from Figure 7 of [11] with permission from Elsevier).

On the other hand, the ethanol content in the permeate was lower and decreased more strongly with the graft content for the copolymers with long grafts (GLX). The phase separation of the long grafts was responsible for their better mobility, which facilitated the mass transfer of the bulky ETBE molecules and, consequently, decreased the membrane selectivity. The copolymer with the highest content of long grafts (GL44), which corresponded to fully segregated graft nano-domains, eventually led to the smallest ethanol permeate content (87.6 wt%), which still remained fairly high for this separation.

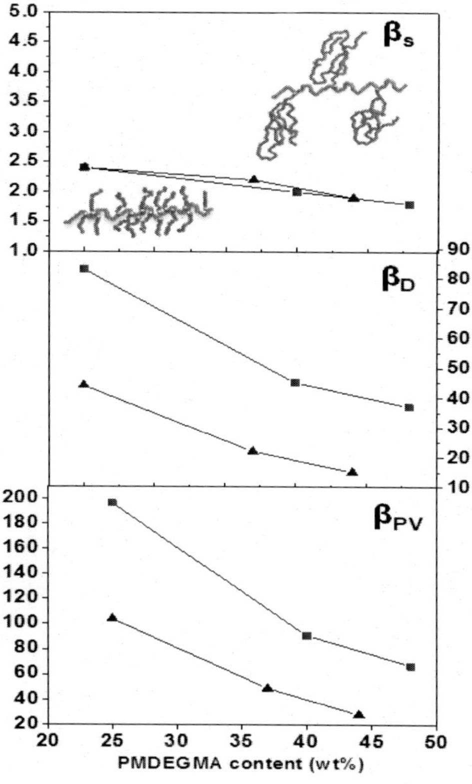

Figure 7. Influence of the copolymer graft content and architecture on diffusion separation factor β_D for the PV separation of the azeotropic mixture ETBE/EtOH at 50°C. (■) Cellulosic copolymers with short PMDEGMA grafts, (▲) Cellulosic copolymers with long PMDEGMA grafts (reproduced from Figure 8 of [11] with permission from Elsevier).

On the basis on the sorption-diffusion model [26], an analysis of the membrane separation factors β_{PV}, β_S, and β_D (for PV, sorption and diffusion, respectively) enabled a further insight into the respective roles played by sorption and diffusion on the global PV selectivity for both series of grafted copolymers [11]. For all of these copolymers, the sorption separation factors were fairly low ($\beta_S < 3$) and they did not vary significantly with the copolymer graft content and architecture. The diffusion separation factors β_D were generally at least one order of magnitude higher than the sorption separation factors β_S. Therefore, the diffusion separation factors were the key contributors to the PV separation factors for both membrane series.

Furthermore, the diffusion separation factors strongly varied with the copolymer graft content and architecture (Figure 7). The copolymers with short grafts were responsible for particularly high diffusion separation factors, which reflected the limitation of the diffusion of the bulky ETBE molecules in these particular materials. On the other hand, the copolymers with the long grafts had significantly decreased diffusion separation factors because their segregated soft grafts favored ETBE diffusion along with that of ethanol.

2.3. Conclusion

New bio-based cellulosic materials with nano-structured architecture were developed from cellulose acetate by a "grafting from" method allowing a good control of the number and length of polymethacrylate soft grafts [25]. The membrane separation properties of the new materials were strongly improved compared to those of cellulose acetate for ETBE purification by PV.

The nano-structured architecture of the new bio-based materials with either long or soft grafts had a strong influence on the membrane properties for this separation [11]. Contrary to the short grafts, the long grafts tended to segregate within the copolymer materials and even formed highly permeable percolating nano-domains for the highest graft content (44 wt%).

Consequently, for the same graft content, the long grafts were responsible for higher flux but lower selectivity compared to short grafts. Therefore, for the design of highly selective bio-based cellulosic membranes, short grafts should be preferred to long grafts for ETBE purification by PV.

3. 100% Bio-Based Cellulosic Materials with Polylactide Grafts for ETBE Purification by Pervaporation

3.1. Strategy for the Synthesis of the New Bio-Based Membrane Materials

In the second strategy for developing new bio-based membranes for ETBE purification by PV, cellulose acetate (CA) was grafted with polylactide (PLA) by "click" chemistry. PLA is a well-known bio-based polyester, which is produced from renewable resources (e.g., corn starch, sugar cane or sugar beet). Nevertheless, membranes based on PLA for the separation of liquids by PV have been very rarely reported so far [12, 27]. The corresponding membranes combined poly(*N*-vinyl pyrrolidone) and PLA in polymer blends and they were described for removing ethanol from cyclohexane or ETBE by PV. For ETBE purification, the best membrane properties corresponded to a polymer blend containing 97 wt% of PLA. The latter membrane had a very high flux (*ca.* 2.7 kg/h m^2) but a moderate selectivity corresponding to an ethanol permeate content of 80 wt% at 30°C [12].

In the new cellulosic membranes developed in our work, the unique properties of CA and PLA were combined in 100% bio-based grafted copolymers (CA-g-PLA) for the highly selective removal of ethanol from ETBE by PV. For preparing the new 100% bio-based membranes, CA was grafted with PLA oligomers by "click" chemistry based on the copper(I) catalyzed 1,3-dipolar cycloaddition of azides and alkynes (CuAAC) (Figure 8). This technique has been widely used for modifying various cellulosic substrates and derivatives for a wide range of applications [28]. According to the main conclusions of our former study on bio-based cellulosic membranes with nano-structured architecture [11] reported in the previous section of this chapter, a low molecular weight (M_n = 1640 g/mol) was chosen for the PLA grafts to ensure high selectivity for the new grafted copolymer membranes.

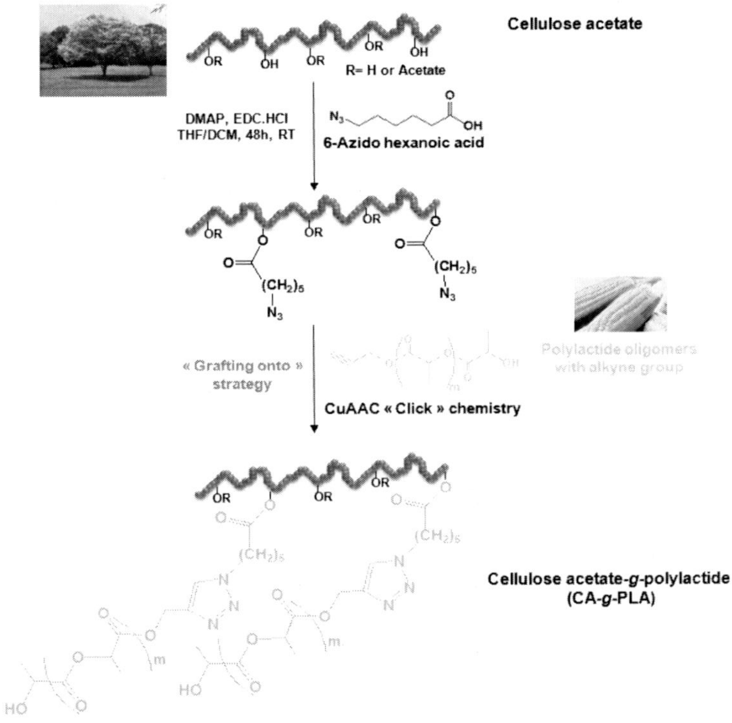

Figure 8. Strategy for the synthesis of 100% bio-based cellulosic materials from cellulose acetate and polylactide for ETBE purification by pervaporation.

According to Figure 8, CA was grafted in two steps by a "grafting onto" method. In the first step, azide side groups for the "click" chemistry were introduced onto CA by esterification with 6-azido hexanoic acid. In the second step, these azido groups were coupled with the terminal alkyne group of short PLA oligomers by CuAAC "click" chemistry. This simple "grafting onto" method enabled to control the number and the length of the PLA grafts. Consequently, the content of the PLA grafts was easily varied from 0 (for virgin CA) to 40.5 wt% in the new cellulosic membranes. The PLA content was limited to below 50 wt% to ensure the mechanical withstanding of the new membranes in the PV operating conditions.

A DSC and synchrotron SAXS analysis revealed that the new materials were not crystalline and that the PLA grafts were dispersed homogeneously at the nano-scale in the new bio-based membranes, as expected for such short grafts. Furthermore, the short PLA grafts acted as very efficient plasticizers for the new cellulosic membrane materials, as shown by the very strong decrease of their glass transition temperature T_g compared to that of virgin CA (T_g = 184°C). As expected, the maximum plasticization was observed for the grafted copolymer containing the highest PLA content (40.5 wt%) with a T_g of 69°C only. This remarkable plasticization effect was responsible for the strong softening of the new membrane materials and for the resulting improvement of their separation properties for ETBE purification by PV compared to virgin CA.

3.2. Structure-Membrane Property Relationships for ETBE Purification by PV

As for our former study on bio-based cellulosic membranes with nano-structured architecture [11], the membrane properties were assessed at 50°C for the PV separation of the target azeotropic mixture ETBE/ethanol (80/20 wt%/wt%) for the new 100% bio-based membranes CA-g-PLA and virgin CA for comparison (Figure 9) [13].

Figure 9 shows that the PV flux of the new membranes strongly increased with their PLA content. The grafting of PLA onto CA was a very efficient way for plasticizing and softening the new membrane materials and eventually inducing a much better mass transfer compared to that in virgin CA. Furthermore, the homogeneous dispersion of the PLA grafts in the new membranes ensured a linear increase of the PV flux as a function of the PLA content. The flux (0.27 kg/h m^2) of the membrane with the highest PLA content (40.5 wt%) was more than one order of magnitude (×12) higher than that of CA.

Nevertheless, the ethanol permeate content C'_{EtOH} remained in the very high range (> 90 wt%) for all the new 100% bio-based membranes (Figure 9). It was extremely high (≥ 99 wt%) for the membrane with the lowest content of grafted PLA (23 wt%), which was also the most rigid material of the grafted copolymers series. In this particular material, the short PLA grafts were highly constrained by the rigid cellulosic chains and these constraints strongly limited the diffusion of the bulky ETBE molecules. When the PLA content increased, the ethanol permeate content decreased, but this decrease was very low, and all of the new membranes were highly selective for ethanol removal from ETBE. For the highest PLA content (40.5 wt%), the short PLA grafts were forming a soft percolating network, which corresponded to a continuous pathway of the PLA grafts inside the new cellulosic membrane. This soft percolating network slightly facilitated the diffusion of the bulky ETBE molecules along with that of ethanol. This resulted in a slightly decreased ethanol permeate content (93 wt%), which still remained in the very high range for ETBE purification by PV.

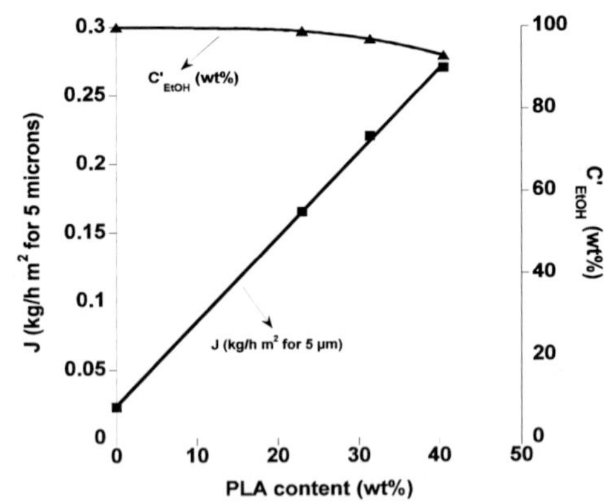

Figure 9. Influence of the polylactide (PLA) content on the membrane flux and the ethanol weight percentage in the permeate for the PV separation of the azeotropic mixture ETBE/EtOH at 50°C for virgin CA and the new 100% bio-based cellulosic membranes CA-g-PLA (reproduced from Figure 8 of [13] with permission from Elsevier).

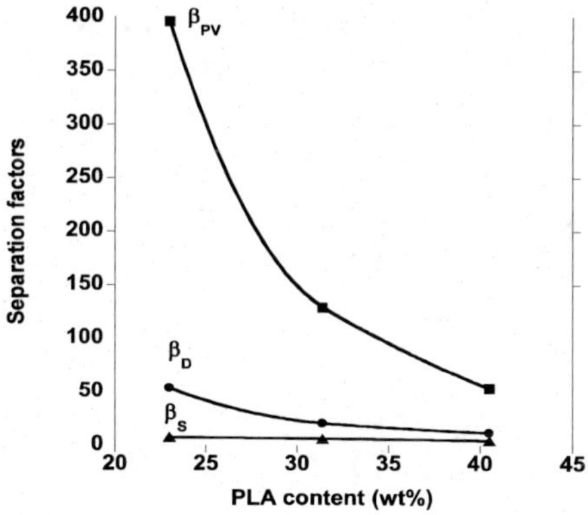

Figure 10. Influence of the polylactide (PLA) content on the membrane separation factors β_{PV}, β_S, β_D related to PV, sorption and diffusion for the PV separation of the azeotropic mixture ETBE/EtOH at 50°C for virgin CA and the new 100% bio-based cellulosic membranes CA-g-PLA (reproduced from Figure 10 of [13] with permission from Elsevier).

Similarly as for the bio-based membranes with nano-structured architecture reported in the previous section of this chapter, an analysis of the membrane separation factors related to PV, sorption and diffusion revealed the contributions of sorption and diffusion to the global PV selectivity (Figure 10). The sorption separation factors β_S of the new 100% bio-based membranes were fairly low ($\beta_S \leq 7.4$) compared to that of virgin CA (β_S = 129), meaning that the sorption step was moderately selective for all the new membranes. The strong decrease in the sorption separation factors of the new grafted copolymers corresponded to the decrease in their hydrogen bonding ability. Nevertheless, the diffusion separation factors β_D of the new membranes were systematically much higher than their sorption separation factors β_D. Therefore, the diffusion step was the main contributor to the selectivity of the global PV mass transfer. The diffusion separation factor β_D decreased from 53.3 for the lowest PLA content (23 wt%) to 10.2 for the highest PLA content (40.5 wt%). For comparison, the diffusion separation factor β_D and the PV separation factor β_{PV} of virgin CA were infinite. Therefore, the corresponding data could not be reported in Figure 10. The grafting of PLA onto CA induced a decrease in the diffusion and PV selectivity for ethanol due to the simultaneous decrease in H-bonding ability and material rigidity. However, the selectivity of the new 100% bio-based membranes still remained in the very high range for ETBE purification by PV as shown by the very high permeate ethanol content (≥ 93 wt%) obtained with all of these membranes.

3.3. Conclusion

New 100% bio-based cellulosic materials were obtained by a simple grafting strategy of short alkyne-PLA oligomers onto azido cellulose acetate by CuAAC "click" chemistry. This strategy enabled to vary the PLA content easily from 0 to 40.5 wt%. The grafting of short PLA oligomers induced a very strong plasticization/softening of the new copolymer materials, finally resulting in a very strong improvement of the membrane flux (up to ×12 compared to virgin cellulose acetate) for ETBE purification by PV [13].

Inspired by the main conclusions of our former study on the bio-based membranes with nano-structured architecture [11], the choice of short PLA grafts was very important to avoid any phase separation in the new 100% bio-based materials. In these new cellulosic membranes, the homogeneous dispersion of the PLA grafts at the nano-scale ensured that the diffusion of the bulky ETBE molecules was strongly hindered due to the strong constraints imposed by the highly rigid cellulosic chains. This particular morphology was eventually responsible for the maintaining of the PV selectivity at a very high level ($C'_{EtOH} \geq 93$ wt%) for all of these new 100% bio-based cellulosic membranes.

4. Bio-Based Cellulosic Materials with Grafted Ionic Liquids for ETBE Purification by Pervaporation

4.1. Introduction

Ionic liquids (ILs) have been widely used as "green solvents" with a lot of specific advantages compared to common organic solvents [29, 30]. Their thermal stability, negligible vapor pressure, and capacity to dissolve a wide range of chemical species have made them attractive for different organic syntheses [30]. For the past ten years, the possibility to control their physical-chemical properties and their hydrophilic/hydrophobic character through the proper choice of their anion and cation has also offered new opportunities for developing innovative materials in the field of separation science for industrial applications [31-38] and bio-products separations [39, 40]. For gas separations, ILs-containing membranes have offered a real breakthrough and several of these new systems have overcome the common permeability/selectivity trade-off with a strong increase in both permeability and selectivity [34, 36, 41-44]. Nevertheless, the separation of liquids by PV still remains particularly challenging for ILs-containing membranes and the related works are much scarce in the literature [14, 24, 39, 45-52].

In the latter case, *organophilic* membranes capable of removing bio-butanols [45-50, 53] or, very rarely, other volatile organic compounds (VOCs) (e.g., ethanol, *iso*-propyl alcohol, toluene etc.) [39, 54, 55] from water, have been mainly reported with supported IL membranes (SILMs) or polymer/IL "composites". In the former works, the IL-containing membranes usually showed improved fluxes with constant or slightly decreased selectivity. More recently, the permanent immobilization of ILs has also been reported for improving membrane stability during PV [53]. In the latter work, the permanent immobilization of a functional triethoxysilane IL was achieved by copolycondensation with a hydroxyl functionalized polydimethylsiloxane and avoided IL dissolution in the feed mixture during butanol recovery.

Hydrophilic membranes containing permanently immobilized ILs for *iso*-propyl alcohol dehydration by PV have also been reported very recently [56]. The covalent grafting of a reactive IL onto cellulose acetate propionate (CAP) led to an excellent improvement in PV separation factor β_{PV} but the PV flux was strongly reduced. Another related strategy for the same application involved the development of CAP/Poly(ionic liquid) (PIL) "composites" based on a polymerizable imidazolium IL [52]. By increasing the PIL content in the new hydrophilic cellulosic membranes, the water flux was greatly improved while the PV separation factor β_{PV} was only slightly reduced, finally leading to an improvement of the Pervaporation Separation Index (PSI) (×2.4) compared to that of virgin CAP.

In this chapter section, we focus on our on-going work on the development of new *organoselective* cellulosic membranes with grafted ILs for ETBE purification by PV [14, 24]. We have already reported that ammonium ILs have a great affinity for ethanol but that these ammonium ILs are usually soluble in the azeotropic mixture ETBE/EtOH [57]. This is an important drawback, which does not allow to develop the corresponding SILMs or polymer/IL "composites" for ETBE purification by PV, because these ILs would be dissolved in the feed mixture during membrane separation. An imidazolium IL has also been reported as an interesting extraction solvent for ETBE purification [58].

We have recently developed an original grafting strategy for the permanent immobilization of ILs containing the same bromide anion (Br^-) and different cations (imidazolium, pyridinium and ammonium) in cellulose acetate membranes for this application. This new strategy avoided IL dissolution during membrane separation and, for the first time, extended the scope of IL-containing membranes to the challenging separation of a purely organic liquid mixture, in which these ILs were soluble. Even more recently, we have used an anion exchange procedure to change the bromide anion by different anions (Tf_2N^-, BF_4^-, AcO^-). Therefore, in this part, the grafting of ILs onto CA will be reported as another very efficient way of inducing CA plasticization. The influence of the IL content and molecular structure will be discussed to reveal the key parameters for the membrane properties. Several of these new bio-based cellulosic membranes with grafted ILs have very improved PV flux (up to almost 20 times that of virgin CA) and permeate ethanol only, corresponding to outstanding infinite PV separation factors β_{PV} for ETBE purification by PV.

4.2. Strategy for the Synthesis of the New Bio-Based Membrane Materials with Grafted Ionic Liquids

The grafting of CA with ILs having the same bromide anion (Br^-) and different cations (imidazolium, pyridinium and ammonium) was achieved by an original strategy in two steps (Figure 11 - Left hand side) [24]. A particular advantage of this simple strategy was to avoid handling copper catalytic species, which had been formerly reported for the grafting of ILs onto cellulose [59, 60]. For these first three series of cellulosic materials grafted with ILs having the same bromide anion and different cations, the bromide anion was initially chosen because it ensured a hydrophilic character for the related ILs and also because of the commercial availability of the corresponding reagent (i.e., 6-bromo hexanoic acid).

Therefore, in the first step, the method of Neises and Steglich [61] was adapted to introduce bromide side groups (with a degree of substitution DS_{Br} = 0.50) onto CA by transesterification of the CA hydroxyl side groups (DS_{OH} = 0.54) with 6-bromo hexanoic acid. In the second step, the different cations (imidazolium, pyridinium and ammonium) were introduced by nucleophilic substitution of the bromide atoms with different nucleophiles (1-methylimidazole, pyridine or *N,N*-diethylmethylamine).

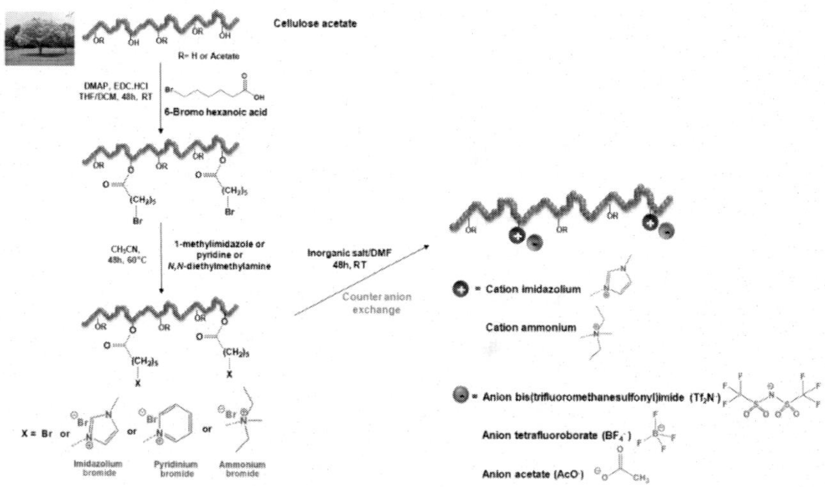

Figure 11. Strategy for the grafting of ionic liquids (ILs) with different cations or anions onto cellulose acetate for ETBE purification by pervaporation.

For these first series of cellulosic materials grafted with different ILs, the IL grafting rate was easily varied by varying the amount of each nucleophile in the reaction medium. As expected, the nucleophile reactivity increased with its nucleophilic character (i.e., pyridine < 1-methylimidazole

<< *N,N*-diethylmethylamine) and a large excess of reactant (8-10 equivalents) was necessary for achieving high grafting rates for the less nucleophilic reactants [24]. Interestingly, the maximum grafting rates for the three series of membranes corresponded to nearly identical maximal amounts of grafted ILs ($W_{IL} \cong 17$ wt%) in the CA membrane materials.

For further improving the membrane separation properties for ETBE purification by PV, the bromide anion of the grafted ILs was then varied by an anion exchange procedure for two CA derivatives grafted with very close amounts ($W_{IL} \cong 17$ wt%) of either imidazolium or ammonium bromide IL (Figure 11 - Right hand side) [14]. These particular cellulosic materials were chosen because they corresponded to the maximum grafting rates and also to the lower and higher fluxes of the first membrane series. Different inorganic salts containing the targeted anions were used for anion exchange in solution in *N,N*-dimethylformamide. The bromide (Br^-) anion was successfully exchanged for the following anions Tf_2N^-, BF_4^- and AcO^-. The anion exchange rate was 100% for both fluorinated anions according to ^{19}F NMR analysis. However, the anion exchange was more difficult for AcO^- due to the poor solubility of the sodium acetate salt used for anion exchange, which limited the anion exchange rate to *ca.* 65%. Nevertheless, eight cellulosic materials grafted with ILs containing imidazolium or ammonium cations and different anions (Br^-, Tf_2N^-, BF_4^-, AcO^-) opened the way for the first systematic study of the anion influence for a series of homologous IL-containing membranes for liquid separation by PV [14].

4.3. Relationships between the Ionic Liquid Structure/Content and the Membrane Properties for ETBE Purification by PV

4.3.1. Influence of the Ionic Liquid Cation

The membrane properties were first assessed at 50°C for the PV separation of the target azeotropic mixture ETBE/ethanol (80/20 wt%/wt%) for membranes made from CA grafted with ILs containing the same bromide anion and different cations (imidazolium, pyridinium and ammonium) [24].

Figure 12 shows that the PV flux was increased nearly 2-fold by CA grafting with 5 wt% of the IL containing the imidazolium cation. However, further increasing the grafting rate of this IL did not lead to any further improvement in the PV flux, which leveled off and remained low for ETBE purification by PV. The PV flux obtained with the highest grafting rate of the IL containing the pyridinium cation was close to that obtained with the imidazolium cation. For this reason, the influence of the IL content on the PV flux was not further investigated for the pyridinium cation. On the other hand, increasing the grafting rate of the IL containing the ammonium cation had a very strong impact of the PV flux, which was increased by a factor up to 8 compared to that of virgin CA. Complementary sorption experiments showed that this strong flux improvement was due to the best membrane swelling obtained with this particular cation [24]. This swelling induced the best plasticization of the grafted cellulosic membrane and the resulting material softening greatly improved membrane permeability.

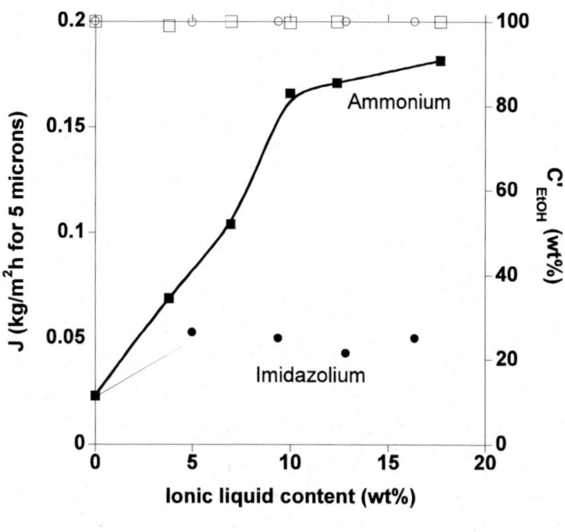

Figure 12. Influence of the grafted IL content in the CA membranes on the PV properties for the separation of the azeotropic mixture ETBE/EtOH at 50°C: Y1 axis : PV flux J for imidazolium (●) and ammonium (■) ILs; Y2 axis: Ethanol weight percentage in the permeate C'_{EtOH} for imidazolium (○) and ammonium (□) ILs (reproduced from Figure 4 of [24] with permission from Elsevier).

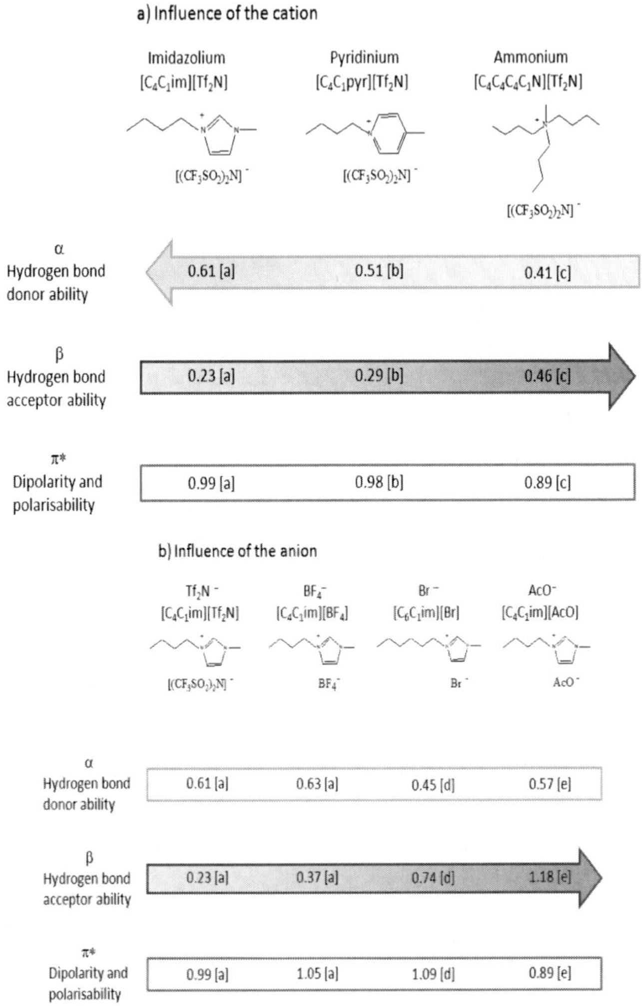

Figure 13. Kamlet-Taft parameters of homologous series of closely related ILs used for the structure-property relationship analysis of the membrane properties for ETBE purification by PV : a) Influence of the IL cation, b) Influence of the IL anion, according to [a] [63], [b] [64], [c] [65], [d] [66], [e] [67].

Furthermore, all of the membranes grafted with ILs containing the same bromide anion and different cations (imidazolium, pyridinium and ammonium) were extremely selective to ethanol with permeate samples containing at least 99 wt% of ethanol. For the maximum IL content ($W_{IL} \cong$

17 wt%), ETBE was not detected anymore and the permeate samples contained ethanol only, which remains exceptional for ETBE purification by PV. The extremely high selectivity of these membranes showed their very strong affinity for ethanol, which is both a strong hydrogen bond donor and a strong hydrogen bond acceptor.

To get a deeper understanding of the very particular behavior of these membranes, the Kamlet-Taft polarity parameters [62] were used to characterize the hydrogen bond donor ability (α) and the hydrogen bond acceptor ability (β) of ethanol and those of a homologous series of ILs closely related to this work in absence of the corresponding data for the grafted ILs [24]. This homologous series enabled a ranking of the different ionic liquids (ILs) in the Kamlet-Taft polarity scale (Figure 13 a)). For ILs containing the same anion and the different cations of interest, the IL hydrogen bond donor ability (α) decreased and the hydrogen bond acceptor ability (β) increased in the following order of the cations: imidazolium, pyridinium and ammonium.

From former works on the ILs Kamlet-Taft parameters, it is well known that the IL cation mainly governs the IL hydrogen bond donor ability (α), which increases in the following order: ammonium < pyridinium < imidazolium [63-65]. Nevertheless, Figure 14 revealed that the PV flux obtained for the ILs differing in their cations varied in *opposite* way compared to this parameter α. Therefore, the IL hydrogen bond donor ability was not the main parameter governing the membrane permeability, although ethanol was a strong hydrogen bonding acceptor ($\beta = 0.77$ [62]).

Figure 14 showed that the PV flux increased with the parameter β, which is known to reflect mainly the hydrogen bond acceptor ability of the anion [63-65]. A better hydrogen bond acceptor ability of the anion improved the IL interactions with ethanol, which is also a strong hydrogen bonding donor ($\alpha = 0.83$ [68]). Therefore, for these first series of membranes with grafted ILs differing in their cations, the IL *anion* capacity to interact with ethanol, which also depended on the cation structure, mainly contributed to the membrane flux.

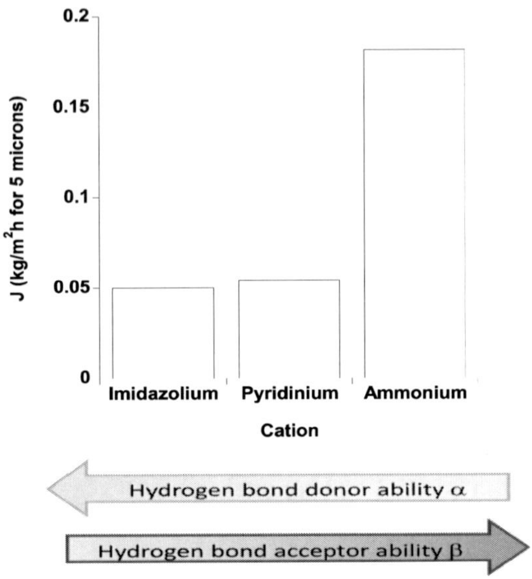

Figure 14. Influence of the IL cation on the PV flux of CA-g-IL membranes for the separation of the azeotropic mixture ETBE/EtOH at 50°C : An analysis based on the IL Kamlet-Taft polarity parameters (adapted from Figure 6 of [24] with permission from Elsevier).

4.3.2. Influence of the Ionic Liquid Anion

After the study of the first series of cellulosic membranes grafted with ILs differing in their cations had revealed that the key contributor for the PV flux was in fact the capacity of the IL anion to interact with ethanol, the influence of the anion was systematically investigated for eight cellulosic membranes grafted with nearly identical amounts ($W_{IL} \cong 17$ wt%) of ILs containing either imidazolium or ammonium cations and different anions (Tf_2N^-, BF_4^-, Br^-, AcO^-) [14].

For the PV of the azeotropic mixture ETBE/EtOH at 50°C with these new cellulosic membranes, the permeate samples contained ethanol only and the membrane selectivity was impressive. Similarly as for the first series of membranes differing in the IL cation, the properties of the new membranes were analyzed based on the Kamlet-Taft polarity parameters of a series of closely related ILs with the different anions of interest (Figure 13 b)). This physical chemical analysis showed that the PV flux of the new membranes also strongly improved with the IL hydrogen bond acceptor ability β in the following order of the anions: Tf_2N^-< BF_4^- < Br^-< AcO^- (Figure 15).

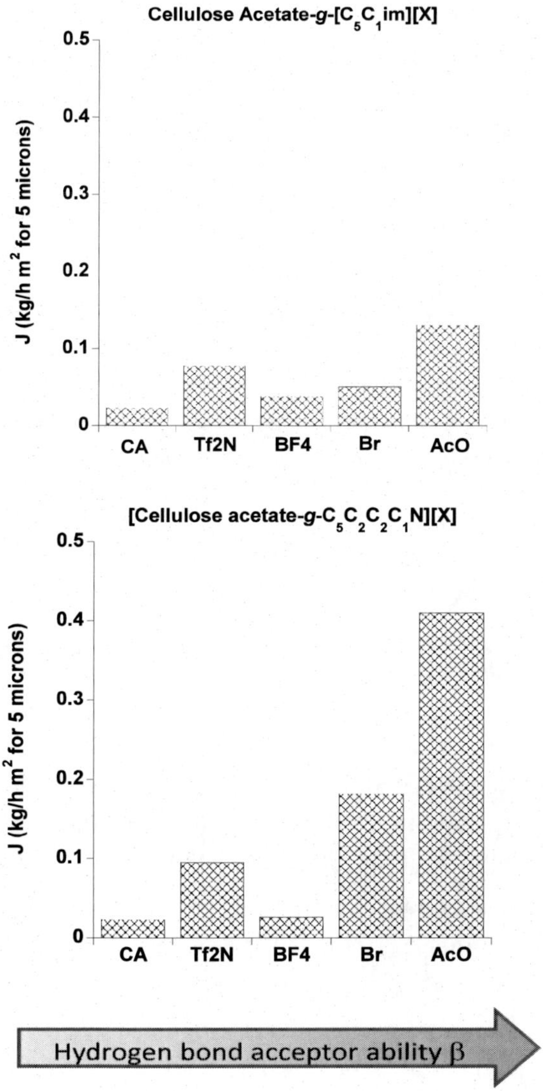

Figure 15. Influence of the IL anion on the PV flux of CA-g-IL membranes for the separation of the azeotropic mixture ETBE/EtOH at 50°C: An analysis based on the IL Kamlet-Taft polarity parameter β characterizing the IL hydrogen bond acceptor ability (reproduced from Figure 5 of [14] with permission from Elsevier).

Nevertheless, the size of the anion also played a significant role on the PV flux, as shown by the further flux improvemement obtained with the biggest anions (Tf_2N^- and AcO^-). For both membrane series with either imidazolium or ammonium cations, increasing the anion size loosened the interactions between the IL cation and anion and eventually improved the interactions with ethanol. This resulted in higher membrane swelling and, consequently, in strong flux improvement. As expected, this improvement was further reinforced for the cation with the biggest size (ammonium) (Figure 15).

Finally, the best membrane properties were achieved with the membrane grafted with an IL containing the ammonium cation and the acetate anion, which combined the best IL hydrogen bond acceptor ability β with big cation and anion sizes. For this particular membrane, the PV flux increased 18-fold compared to virgin CA and the permeate contained ethanol only, corresponding to an outstanding selectivity and an infinite PV separation factor β_{PV} for ETBE purification by PV.

4.4. Conclusion

A new strategy was successfully developed for grafting ionic liquids (ILs) onto cellulose acetate (CA) in order to avoid ILs dissolution by the feed mixture during ETBE purification by PV. For the first time, this grafting strategy enabled to use IL-containing membranes for separating a purely organic feed mixture, in which these ILs were soluble.

An analysis based on the Kamlet-Taft polarity parameters showed that the membrane properties were mainly governed by the ability of the *anion* to interact with ethanol by hydrogen bonding. In addition, increasing the IL cation and anion sizes also improved the membrane properties by weakening the IL cation/anion interactions and favoring those with ethanol. Consequently, combining the best IL hydrogen bond acceptor ability β with big cation and anion sizes finally led to the best membrane properties for the bio-based cellulosic material CA-g-[ammonium][acetate].

CONCLUSION

High performance bio-based membranes for ETBE purification by pervaporation have been obtained by different controlled grafting strategies of a major cellulosic derivative (cellulose acetate (CA)). Various types of grafts (i.e., polymethacrylate oligomers containing oxyethylene monomer units, bio-based polylactide oligomers and ionic liquids with different cations and anions) were successfully considered to plasticize the cellulosic membranes and improve their flux while maintaining a very high selectivity for the targeted application.

First, polymethacrylate grafts of different lengths revealed the importance of the copolymer architecture and graft nano-structuration on the membrane properties. Long grafts induced a strong graft nano-structuration, which greatly improved the membrane flux but the selectivity was decreased significantly compared to that of CA. On the other hand, short grafts were homogeneously dispersed and highly constrained by the rigid cellulosic chains in the new grafted cellulosic copolymers. By restricting the mobility of the bulky ETBE molecules, this particular copolymer architecture led to the best membrane properties with an improved flux (up to ×48 compared to CA) and a very high selectivity ($C'_{EtOH} = 94.4$ wt%) for ETBE purification by pervaporation.

Secondly, the grafting of CA with short bio-based polylactide (PLA) oligomers led to 100% bio-based membranes, both of the components of the grafted copolymers coming from wood and corn, respectively. The short polylactide grafts induced a very strong plasticization of the cellulosic membranes and, consequently, a great improvement of the membrane flux (up to ×12 compared to CA). Similarly as what had been observed with the polymethacrylate grafts, the homogeneous dispersion of the short polylactide grafts in the new cellulosic materials enabled to maintain the membrane selectivity at a very high level ($C'_{EtOH} > 90$ wt%) for the targeted application.

Finally, the grafting of CA with ionic liquids (ILs) containing the same bromide anion and different cations (imidazolium, pyridinium and ammonium) was another efficient way for improving the membrane properties for

ETBE purification by PV. The membrane flux increased in the following order of the cation: imidazolium < pyridinium << ammonium. Nevertheless, for these first series of cellulosic materials grafted with ILs differing in their cation, an analysis based on the Kamlet-Taft polarity parameters showed that the membrane properties were in fact mainly governed by the ability of the *anion* to interact with ethanol by hydrogen bonding. This was further confirmed by the systematic investigation of the anion influence in a second series of cellulosic membranes grafted with ionic liquids having either imidazolium or ammonium cations, and different anions (Tf_2N^-, BF_4^-, Br^-, AcO^-). The size of the anion and cation also played a strong role in the membrane performance. Increasing their size decreased the intensity of the cation/anion interactions and favored the interactions with ethanol by hydrogen bonding. Consequently, the best membrane flux (up to almost 20 times that of CA) was obtained by combining the ammonium cation with the acetate anion. Furthermore, all the membranes with the maximum IL content (*ca.* 17 wt%) permeated ethanol only (C'_{EtOH} = 100 wt%), corresponding to an outstanding selectivity for ETBE purification by PV.

In the future, these controlled grafting strategies could be applied to different polysaccharides to further extend the scope of the present work to a wide range of new high performance bio-based membranes for ETBE purification. They could also be combined together in innovative bio-based materials, for example by associating ionic liquids and polymer grafts of various natures including poly(ionic liquid)s (PILs). Last but not least, these grafting strategies could also offer new solutions for separating other challenging liquid mixtures by PV. In particular, the future prospects include solvent and biofuel dehydration processes, which are currently being the subject for a renewed interest for energy saving and sustainable development.

Acknowledgments

The authors would like to thank the different funding agencies having supported their work on the development of new bio-based membranes for

ETBE purification by pervaporation over the past decade: Agence de l'environnement et de la maîtrise de l'énergie (ADEME), Centre National de la Recherche Scientifique (CNRS), Université de Lorraine (UL) and the ELEMENT Erasmus Mundus Programme from the European Union (EU). They also would like to gratefully acknowledge the contributions of the different following researchers to the development of the new membrane materials or to their characterization: Robert CLEMENT, Laurent DAVID, Marlène DRESCH, Pierre LOCHON, Cécile NOUVEL, Andrea RANZANI DA COSTA, Jean-Luc SIX and Guillaume SUDRE.

REFERENCES

[1] Noureddini, H. *ACS Symp. Ser.* 2002, 799, 107-124.

[2] Yee, K. F.; Mohamed, A. R.; Tan, S. H. *Renewable Sustainable Energy Rev.* 2013, 22, 604-620.

[3] Wang, M.; Arnal-Herault, C.; Rousseau, C.; Palenzuela, A. l.; Babin, J.; David, L.; Jonquieres, A. *J. Membr. Sci.* 2014, 469, 31-42.

[4] Weber de Menezes, E.; Cataluna, R. *Fuel Process. Technol.* 2008, 89, 1148-1152.

[5] Chen, M. S.; Eng, R. M.; Glazer, J. L.; Wensley, C. G., *Pervaporation process for separating alcohols from ethers*, US patent 7,774,365, 1988.

[6] Streicher, C., Institut Francais du Pétrole, *Ethyl tertio-butyl ether purification process combining a membrane method and distillation*, US Patent 5,607,557, 1997.

[7] Drioli, E.; Giorno, L.; Eds; *Comprehensive Membrane Science and Engineering*, Elsevier Science: Kidlington (UK), 2010.

[8] Jonquieres, A.; Arnal-Herault, C.; Babin, B. In *Encyclopedia of Membrane Science and Technology*; Hoek, E. M. V. Tarabara, V. V.; Eds.; Wiley: Hoboken, New Jersey, 2013; Vol. 3 pp. 1533-1559.

[9] Norkobilov, A.; Gorri, D.; Ortiz, I. *Chem. Eng. Process.* 2017, 122, 434-446.

[10] Jonquieres, A.; Clement, R.; Lochon, P.; Dresch, M.; Chrétien, B. *J. Membr. Sci.* 2002, 206, 87-117.

[11] Billy, M.; Ranzani Da Costa, A.; Lochon, P.; Clement, R.; Dresch, M.; Jonquieres, A. *J. Membr. Sci.* 2010, 348, 389-396.

[12] Zereshki, S.; Figoli, A.; Madaeni, S. S.; Galiano, F.; Drioli, E. *J. Membr. Sci.* 2011, 373, 29-35.

[13] Hassan Hassan Abdellatif, F.; Babin, J.; Arnal-Herault, C.; Nouvel, C.; Six, J. L.; Jonquieres, A. *J. Membr. Sci.* 2017, 524, 449-459.

[14] Hassan Hassan Abdellatif, F.; Babin, J.; Arnal-Herault, C.; David, L.; Jonquieres, A. *Carbohydr. Polym.* 2018, 196, 176-186.

[15] Nguyen, Q. T.; Leger, C.; Billard, P.; Lochon, P. *Polym. Adv. Technol.* 1997, 8, 487-495.

[16] Shao, P.; Huang, R. Y. M. *J. Membr. Sci.* 2007, 287, 162-179.

[17] Uragami, T. In *Polysaccharides (2nd Edition)*; Dumitriu, S.; Ed.; Marcel Dekker, Inc. New York, 2005, pp. 1087-1122.

[18] Luo, G. S.; Niang, M.; Schaetzel, P. *J. Membr. Sci.* 1997, 125, 237-244.

[19] Nguyen, Q. T.; Clement, R.; Noezar, I.; Lochon, P. *Sep. Purif. Technol.* 1998, 13, 237-245.

[20] Luo, G. S.; Niang, M.; Schaetzel, P. *J. Appl. Polym. Sci.* 1997, 66, 1631-1638.

[21] Luo, G. S.; Niang, M.; Schaetzel, P. *Sep. Sci. Technol.* 1999, 34, 391-401.

[22] Léger, C.; Nguyen, Q. T.; Néel, J. *Macromolecules* 1995, 28, 143-151.

[23] Billard, P.; Nguyen, Q. T.; Leger, C.; Clément, R. *Sep. Purif. Technol.* 1998, 14, 221-232.

[24] Hassan Hassan Abdellatif, F.; Babin, J.; Arnal-Herault, C.; David, L.; Jonquieres, A. *Carbohydr. Polym.* 2015, 147, 313-322.

[25] Billy, M.; Ranzani Da Costa, A.; Lochon, P.; Dresch, M.; Clément, R.; Etienne, S.; Hiver, J. M.; David, L.; Jonquieres, A. *Eur. Polym. J.* 2010, 46, 944-957.

[26] Wijmans, J. G.; Baker, R. W. *J. Membr. Sci.* 1995, 107, 1-21.

[27] Zereshki, S.; Figoli, A.; Madaeni, S. S.; Simone, S.; Jansen, J. C.; Esmailinezhad, M.; Drioli, E. *J. Membr. Sci.* 2010, 362, 105-112.

[28] Hassan Hassan Abdellatif, F.; Babin, J.; Arnal-Herault, C.; Jonquieres, A. In *Cellulose-based Graft Copolymers: Structure and Chemistry*; Thakur, V. K.; Ed.; CRC Press Taylor & Francis Publisher: Boca Raton, Florida, 2015, pp. 569-597.

[29] Rogers, R.; Seddon, K.; Eds; *Ionic liquids. Industrial applications to green chemistry*, ACS Symposium Series 818, American Chemical Society: Washington, DC, 2002.

[30] Wasserscheid, P.; Welton, T. *Ionic liquids in Synthesis*; 2nd ed; Wiley-VCH: Weinheim, 2008.

[31] Chen, J.; Rodríguez, H.; Rogers, R. D. *Separ. Sci. Technol.* 2012, 47, 167-168.

[32] Green, M. D.; Long, T. E. *J. Macromol. Sci., Polym. Rev.* 2009, 49, 291–314.

[33] Han, D.; Row, A. K. H. *Molecules* 2010, 15, 2405-2426.

[34] Lozano, L. J.; Godinez, C.; de los Rios, A. P.; Hernandez-Fernandez, F. J.; Sanchez-Segado, S.; Alguacil, F. J. *J. Membr. Sci.* 2011, 376, 1-14.

[35] Lu, J.; Yan, F.; Texter, J. *Progr. Polym. Sci.* 2009, 34, 431-448.

[36] Noble, R. D.; Gin, D. L. *J. Membr. Sci.* 2011, 369, 1-4.

[37] Scovazzo, P.; Visser, A. E.; Davis, J., H.; Rogers, R., D.; Koval, C., A.; DuBois, D., L.; Noble, R. D. *ACS Symp. Ser.* 2002, 818 (Ionic liquids), 69-87.

[38] Rynkowska, E.; Fatyeyeva, K.; Kujawski, W. *Rev. Chem. Eng.* 2018, 34, 341-363.

[39] Izak, P.; Friess, K.; Sipek, M. In *Handbook of membrane research: Properties, performances and applications*; Gorley, S. V.; Ed.; Nova Science Publishers: New York, 2010, pp. 387-402.

[40] Roosen, C.; Müller, P.; Lasse, G. *Appl. Microbiol. Biotechnol.* 2008, 81, 607-614.

[41] Close, J. J.; Farmer, K.; Moganty, S. S.; Baltus, R. E. *J. Membr. Sci.* 2012, 390, 201-210.

[42] Hanioka, S.; Maruyama, T.; Sotani, T.; Teramoto, M.; Matsuyama, H.; Nakashima, K.; Hanaki, M.; Kubota, F.; Goto, M. *J. Membr. Sci.* 2008, 314, 1-4.

[43] Hojniak, S. D.; Silverwood, I. P.; Khan, A. L.; Vankelecom, I. F. J.; Dehaen, W.; Kazarian, S. G.; Binnemans, K. *J. Phys. Chem. B* 2014, 118, 7440-7449.

[44] Scovazzo, P. *J. Membr. Sci.* 2009, 343, 199-211.

[45] Cascon, H. R.; Choudhari, S. K. *J. Membr. Sci.* 2013, 429, 214-224.

[46] Fadeev, G.; Meagher, M. M. *Chem. Comm.* 2001, 3, 295-296.

[47] Heitmann, S.; Krings, J.; Kreis, P.; Lennert, A.; Pitner, W. R.; Garak, A.; Schulte, M. M. *Sep. Purif. Technol.* 2012, 97, 108-114.

[48] Izak, P.; Ruth, W.; Fei, Z.; Dyson, P. J.; Kragl, U. *Chem. Eng. J.* 2008, 139, 318-321.

[49] Matsumoto, M.; Murakami, Y.; Kondo, K. *Solv. Extr. Res. Dev.* 2011, 18, 75-83.

[50] Plaza, A.; Merlet, G.; Hasanoglu, A.; Isaacs, M.; Sanchez, J.; Romero, J. *J. Membr. Sci.* 2013, 444, 201-212.

[51] Yahaya, G. O.; Hamad, F.; Bahamdan, A.; Tammana, V. V. R.; Hamad, E. Z. *Fuel Process. Technol.* 2013, 113, 123-129.

[52] Rynkowska, E.; Dzieskowski, K.; Lancien, A.; Fatyeyeva, K.; Szymczyk, A.; Kujawa, J.; Koter, S.; Marais, S.; Wolan, A.; Kujawski, W. *J. Membr. Sci.* 2017, 544, 243-251.

[53] Mai, N. L.; Kim, S. H.; Ha, S. H.; Shin, H. S.; Koo, Y. M. *Korean J. Chem. Eng.* 2013, 30, 1804-1809.

[54] Kohoutova, M.; Sikora, A.; Hovorka, Å.; Randova, A.; Schauer, J.; Tiama, M.; Setničkova, K.; Petričkovič, R.; Guernik, S.; Greenspoon, N.; Izak, P. *Eur. Polym. J.* 2009, 45, 813-819.

[55] Takekoshi, K.; Miyata, T.; Uragami, T., *Proceedings of the Pacificchem 2010 Conference,* December 15-20, 2010, Honolulu, Hawai, pp. MATNANO-385.

[56] Rynkowska, E.; Fatyeyeva, K.; Kujawa, J.; Dzieszkowski, K.; Wolan, A.; Kujawski, W. *Polymers* 2018, 10, 11-18.

[57] Jonquieres, A.; Awkal, M.; Clément, R.; Lochon, P. In *Polyimides and High Temperature Polymers: Synthesis, Characterization and Applications*; Mittal, K. L.; Ed.; Koninklijke Brill NV Edition: Leiden, The Netherlands, 2009; Vol. 5, pp. 339-351.

[58] Arce, A.; Rodriguez, H.; Soto, A. *Green Chem.* 2007, 9, 247-253.

[59] Gonsior, N.; Ritter, H. *Macromol. Chem. Phys.* 2011, 212, 2633-2640.

[60] Xu, W. Z.; Gao, G.; Kadla, J. F. *Cellulose* 2013, 20, 1187-1199.

[61] Neises, B.; Steglich, W. *Angew. Chem., Int. Ed.* 1978, 17, 522–524.

[62] Kamlet, M. J.; Taft, R. W. *J. Am. Chem. Soc.* 1976, 98, 377-383.

[63] Ab Rani, M. A.; Brant, A.; Crowhurst, L.; Dolan, A.; Lui, M.; Hassan, N. H.; Hallett, J. P.; Hunt, P. A.; Niedermeyer, H.; Perez-Arlandis, J. M.; Schrems, M.; Welton, T.; Wilding, R. *Phys. Chem. Chem. Phys.* 2011, 13, 16831-16840.

[64] Lee, J. M.; Ruckes, S.; Prausnitz, J. M. *J. Phys. Chem. B* 2008, 112, 1473-1476.

[65] Spange, S.; Lungwitz, R.; Schade, A. *J. Molec. Liquids* 2014, 192, 137-143.

[66] Jelicic, A.; Garcia, N.; Lohmannsroben, H. G.; Beuermann, S. *Macromolecules* 2009, 42, 8801-8808.

[67] Doherty, T. V.; Mora-Pale, M.; Foley, S. E.; Linhardt, R. J.; Dordick, J. S. *Green Chem.* 2010, 12, 1967-1975.

[68] Kamlet, M. J.; Abboud, J. L. M.; Abraham, M. H.; Taft, R. W. *J. Org. Chem.* 1983, 48, 2877-2887.

INDEX

A

B

C

D

E

F

G

H

I

K

M

N

O

P

Q

R

S

T

V

W

X

Z